# DEVELOPMENTS
## IN
# APPLIED
# SPECTROSCOPY

### Volume 1

W0230519

*A Publication of the Society for Applied Spectroscopy, Chicago Section*

# DEVELOPMENTS IN APPLIED SPECTROSCOPY

## Volume 1

*edited by*

## W. D. Ashby

Proceedings of the
Twelfth Annual Symposium on Spectroscopy
Held in Chicago, Illinois
May 15-18, 1961

*Distributed by*

℗

**PLENUM PRESS**
NEW YORK
1962

ISBN-13: 978-1-4684-7623-1     e-ISBN-13: 978-1-4684-7621-7
DOI: 10.1007/ 978-1-4684-7621-7

Library of Congress Catalog Card No. 61-17720

# Foreword

On May 15-18, 1961, the 12th Annual Symposium on Spectroscopy was held in Chicago. Over the twelve-year history of this meeting, it has continually grown and now ranks as one of the major technical meetings in the field of spectroscopy.

The scope of the program itself represents a balance between research applications and control applications, between applied and the more fundamental aspects of spectroscopy. Papers are presented each year in the specialty areas of X-ray, ultraviolet, visible, infrared, Raman, EPR, and NMR spectroscopy.

In many instances over the years, excellent work has been reported at this symposium and no further publication was made. These reports were then essentially lost for further reference. It is the purpose of this publication to provide a source of reference for the papers presented at the 12th Annual Symposium.

This first attempt at publishing a proceedings does not include the entire program. Several papers could not be given company clearance for publication, and several more were intended for verbal presentation only and were never written for publication.

I, as Coordinator of the Symposium, would like to express my gratitude to the Symposium Committee, John Ferraro, Elwin Davis, Joseph Ziomek, John Kapetan, L. S. Gray, Jr., Russell J. Hansen, J. A. Sheinkop, L. V. Azaroff, and Carl Moore, whose diligence and labor resulted in a truly fine symposium, and subsequently made possible this publication.

William D. Ashby

# Contents

*Asterisks (\*) indicate that only an abstract appears in this volume*

## X-RAY SPECTROSCOPY

## ULTRAVIOLET AND VISIBLE SPECTROSCOPY

## INFRARED AND RAMAN SPECTROSCOPY
## AND GAS CHROMATOGRAPHY

# X-ray Spectroscopy

# X-ray Spectrometric Determination of Thorium in Purified Uranium Materials

G. R. Blank and H. A. Heller

National Lead Company of Ohio
Cincinnati, Ohio

An X-ray spectrometric method for the determination of thorium in a variety of purified uranium materials is presented. The thorium is separated from the uranium prior to X-ray determination by a 2-thenoyltrifluoro-acetone–xylene extraction. An acid solution of the reextracted thorium is irradiated in a 100-kv X-ray spectrograph after the addition of strontium as an internal standard. A statistical study indicates a limit of error of ±7.7 ppm over the concentration range from 100 to 1000 ppm of thorium.

## INTRODUCTION

In the production of purified uranium materials, thorium is a common and important contaminant whose concentration must be routinely determined. Prior to approximately two years ago, thorium had been determined at this laboratory by chemical (gravimetric and colorimetric) methods [1-3].

A satisfactory X-ray spectrometric method for the determination of thorium in uranium concentrates [4] was then developed and has since been in routine use. This method, based on the irradiation of dry-powder sample preparations, has been very satisfactory for the determination of thorium in concentrations greater than 1000 ppm.

A need existed, however, for an X-ray method capable of detecting and quantitatively determining thorium in the concentration range from 1000 ppm down to at least 100 ppm, because this is the

This work was performed under Atomic Energy Commission Contract No. AT(30-1)-1156.

3

Fig. 1. "Goniometer end" of spectrometer, showing control panel and electronic circuit panel.

range of thorium concentrations normally encountered in purified
uranium materials.

## INSTRUMENTATION

The instrumentation, used both in the preliminary investigations
and in the subsequent routine analytical work, consists of a Norelco
100-kv X-ray spectrograph and the usual accessory equipment. General views of the major assemblies are shown in Figs. 1 and 2.

The design of this instrument represents the so-called "inverted"
geometry, in which the X-ray tube is mounted beneath the sample
and the X-ray beam is directed upward to irradiate the bottom of
the sample. Figure 3 shows the relationship and functions of the
major components.

Excitation is provided by a type FA-100 tungsten-target X-ray
tube. The spectral dispersing element used in this work was a plane,
single crystal of lithium fluoride. The optical path was air throughout and the X-ray detector was a multiplier phototube fitted with

a sodium iodide (thallium-activated) scintillation crystal. The conventional Norelco circuit panel was used for measuring intensities.

## EXPERIMENTAL

Preliminary experimental work was divided into two phases: (1) establishment of optimum conditions for the X-ray determination of thorium and (2) development, or adaptation, of a suitable method for the chemical separation of thorium from uranium. In actual analytical practice, the chemical separation would necessarily precede the X-ray determination. In the preliminary experimental work, however, the conditions of the X-ray determination were established first in order to have a means for evaluating the efficiency of the proposed chemical separation.

### X-ray Determination

A decision was made at the outset to base the X-ray determination on the irradiation of a solution containing the separated thorium,

Fig. 2. Front view of spectrometer, showing sample changer, high-voltage controls, *etc.*

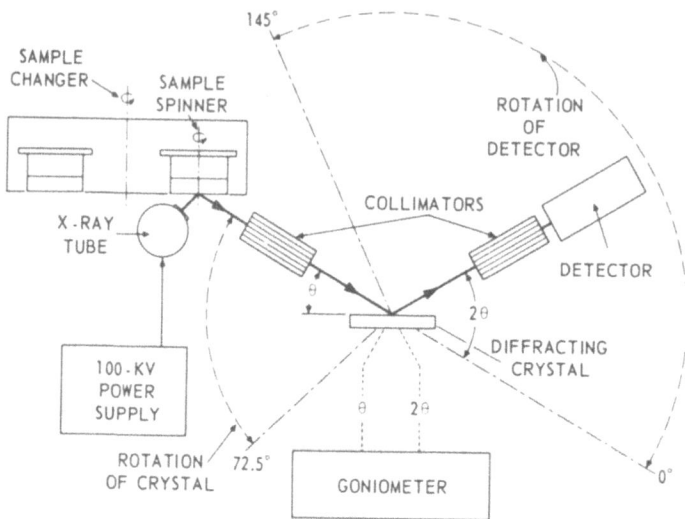

Fig. 3. Geometry of 100-kv X-ray spectrometer.

rather than on a (necessarily) small solid residue. Previous experience indicated that the inherent homogeneity of solutions presents considerable advantages from the standpoints of precision and the ease with which an internal standard element may be added to the sample preparation.

**Liquid-Sample Cell.** The type of liquid-sample holder which has been adopted at this laboratory, and which has been in routine use for some time in other X-ray spectrometric procedures [5,6], consists of a commercially available, cup-shaped polyethylene cell, $1^3/_4$ in. in diameter by $^1/_2$ in. deep (Caplug, Type EC-28).* A Mylar† window, 0.0005 in. thick, is placed over the open end of the cell and secured by means of a tight-fitting polyethylene band approximately $^1/_2$ in. wide. This provides a rugged, liquid-tight, and reasonably flat X-ray "window." It was this type of liquid-sample cell which was used in the present work.

**Concentration Range.** It appeared desirable to determine the thorium in nitrate rather than in chloride solutions in order to secure the advantages of a lower-atomic-number matrix. These advantages

---

*Caplug—Dust and shipping protectors, manufactured by Protective Closures Co., Inc., Buffalo 23, New York.

†Mylar—Polyester film, manufactured by E. I. du Pont de Nemours and Company, Inc., Film Department, Wilmington, Delaware.

include lower X-ray absorption and somewhat lower background intensities.

An investigation was made to determine the range of thorium concentrations in the X-ray specimen solution which provide the best reproducibility and accuracy. A series of thorium nitrate solutions were prepared to contain thorium in concentrations ranging from 10 to 5000 $\mu$g/ml. Ten-milliliter portions of each of these solutions were irradiated in the X-ray spectrometer (60 kv at 25 ma) and the intensities of the Th $L_{\alpha_1}$ emission were recorded. It was found that although concentrations as low as 40 $\mu$g Th/ml could be detected under these conditions, an analytical range of 100 to 1000 $\mu$g Th/ml was adopted because the reproducibility obtained in this range was better than at concentrations lower than 100 $\mu$g Th/ml.

**Internal Standard.** In order to obtain the precision and accuracy required, the use of an added internal standard element was necessary. The use of a properly selected internal standard could be expected to compensate for the following possible sources of error: (1) interelement effects which, in this case, could arise from the incomplete separation of the very large amounts of uranium in the original sample materials; (2) changes in volume (and therefore in concentration) of sample preparations resulting from evaporation and/or temperature changes; (3) small transfer losses or volumetric errors; and (4) variations in irradiation geometry, *e.g.*, variations in sample placement and/or lack of flatness of cell windows.

The use of rubidium as an added internal standard was considered because of the proximity of the energy of the Rb $K_{\alpha}$ emission (13.4 kev) to that of Th $L_{\alpha_1}$ (13.0 kev). The use of rubidium, however, would require a complete absence of uranium from the final sample solutions because the Rb $K_{\alpha}$ peak cannot be resolved from the U $L_{\alpha_2}$ peak using a lithium fluoride crystal and a 0.005-in. collimator.

Bromine, as sodium bromide, is successfully used at this laboratory as an internal standard for the determination of thorium in dry-powder preparations of uranium concentrates [4]. Brief tests showed, however, that bromides are not stable in nitric acid solution upon heating and evaporation (it was expected that it would be necessary to reduce the volume of the thorium-bearing aqueous phase to approximately 10 ml by evaporation after the chemical separation).

The use of strontium as an internal standard for the determination of thorium in solutions containing small amounts of uranium has been reported by Pish and Huffman [7]. Although the energy relationship of the Sr $K_\alpha$ emission (14.2 kev) to the Th $L_{\alpha_1}$ emission (13.0 kev) is not as close as that provided by Rb $K_\alpha$ (13.4 kev), the Sr $K_\alpha$ peak is free of spectral interference from uranium. Strontium, in the form of a standard solution of $SrCO_3$ in dilute nitric acid was, therefore, chosen for use as an added internal standard.

Preliminary analytical curves were prepared by scaling a series of standard thorium solutions (100 to 1000 $\mu g$ Th/ml) to which strontium had been added in varying amounts. A concentration of 175 $\mu g$ Sr/ml was chosen because it provided a "net" intensity ratio (background intensity subtracted from both Th $L_{\alpha_1}$ and Sr $K_\alpha$) of unity at the midpoint (500 $\mu g$ Th/ml) of the proposed analytical range.

A test showed that (Th $L_{\alpha_1}$-bkg)/(Sr $K_\alpha$-bkg), the net intensity ratio, was not appreciably affected by the presence of 1000 $\mu g$ U/ml in the X-ray sample solutions, and that the presence of as much as 2000 $\mu g$ U/ml caused a change of only 1% in the intensity ratio.

**Standards.** A series of gravimetrically standardized solutions covering the range from approximately 48 to 962 $\mu g$ Th/ml was prepared. These were used to establish the final analytical curve after the addition of a 10–ml portion of strontium nitrate solution (1.75 mg Sr/ml) to each standard followed by dilution to 100.0 ml.

**Excitation.** Operating conditions of 60 kv at 25 ma for the tungsten-target X-ray tube were adopted after it was determined that these conditions provided efficient excitation with reasonable intensities throughout the desired concentration range. A test showed that deviations of ±5 kv in the high-voltage supply did not affect the net intensity ratios.

**Specimen Size.** A test was conducted to determine the minimum volume of sample solution in the polyethylene sample cells which would produce absolute intensities not affected by small changes in sample volume. It was found that this requirement was satisfied by a specimen volume of 10 ml, corresponding to a depth of approximately 6 mm in the $1^3/_4$-in. cells. It was noted that, although absolute line intensities were much lower, the net intensity ratios were unchanged when volumes as small as 5 ml were used. In order to maintain good counting statistics, however, a nominal specimen volume of 10 ml was adopted for routine use.

### Chemical Separation

The following methods for chemically separating small amounts of thorium from a uranium matrix were investigated:

1. Precipitation of thorium as the fluoride.
2. Selective extraction of thorium into a solution of thenoyltrifluoroacetone (in xylene) followed by reextraction into dilute nitric acid.

**Fluoride Precipitation.** Thorium may be precipitated as the fluoride from hydrochloric acid solution by the addition of hydrofluoric acid. The complete precipitation of the amounts of thorium (1 to 10 mg) involved in the proposed X-ray method would require as long as three hours.

An "accelerated" method [8] for the precipitation of $ThF_4$ by centrifugation was investigated. The volume limitations of the only centrifuge available for these investigations made it necessary to exceed the recommended concentrations of both thorium and uranium. Under these conditions, the recovery of thorium was incomplete. No further efforts to separate thorium as the fluoride were made.

**TTA Extraction.** Thorium is separated from uranium by extraction into a 0.5$M$ solution of TTA (in xylene) as a preliminary step in a number of analytical methods [1-3]. This extraction is made from a nitric acid solution at a controlled pH, after which the organic phase is washed with dilute (0.1 to 0.2$M$) nitric acid to remove any uranium which might have been carried into the organic phase. The thorium is then reextracted (stripped) from the organic phase into 2$M$ nitric acid. Tracer studies made by Laux [3], using $Th^{230}$, indicated a substantially quantitative recovery of thorium in this separation.

Slight modifications of Laux's procedure were made in order to adapt the TTA extraction to the requirements of the X-ray determination. The modified procedure is described below.

A uranium sample (uranium metal, $U_3O_8$, $UO_3$, *etc.*) of sufficient size to contain 1 to 10 mg of thorium is dissolved in nitric acid and diluted to a volume such that the uranium concentration does not exceed 34 g U/liter. The pH of the solution is adjusted to 1.0 ± 0.1 and the solution is equilibrated with a 0.5$M$ solution of TTA in xylene by shaking in a separatory funnel. A volume of TTA–xylene equal to one-third the volume of the aqueous phase is used. The or-

ganic phase is then washed with a portion of $0.2M$ $HNO_3$, equal in
volume to twice that of the organic phase. The thorium is then
stripped from the organic phase by shaking with two successive portions of $2.0M$ $HNO_3$, the volume of each portion being equal to one-
fourth that of the organic phase. The two portions of $2.0M$ $HNO_3$
(now containing the thorium) are combined and 1.00 ml of a standard
strontium nitrate solution (1.75 mg Sr/ml) is added. The solution
is then evaporated to a volume slightly less than 10 ml. The volume
of the cooled solution is adjusted to approximately 10 ml and analyzed on the X-ray spectrometer.

The X-ray specimen usually contains 300 to 500 $\mu$g U/ml, which
does not, however, interfere with the X-ray determination of
thorium. Tests have shown that this amount of uranium contamination can be reduced to less than 50 $\mu$g U/ml by washing the
organic phase twice, rather than once, with $0.2M$ $HNO_3$. However,
inasmuch as the use of the strontium internal standard apparently
compensates for the presence of as much as 1000 $\mu$g U/ml, the second wash is not used in the routine procedure.

**Evaluation of Thorium Recovery.** Two independent studies were
made to check the efficiency of the modified separation procedure.
In the first test, known amounts of thorium were added to each of
three purified (thorium-free) uranyl nitrate solutions. The thorium
was then extracted and recovered from each solution in accordance
with the modified separation procedure described above. The thorium recoveries were determined by referring the net (X-ray) intensity ratios obtained to an analytical curve which had been prepared
from standardized aqueous thorium nitrate solutions, to each of
which the customary amount of strontium internal standard was
added. An average thorium recovery of 97.9% was found.

In the second test, uranyl nitrate solutions were spiked with
$Th^{234}$ tracer. These solutions were then carried through the modified separation procedure, as described above, and the thorium-
tracer activities were radiometrically determined. In these tests, an
average thorium recovery of 98.1% was found.

## SUMMARY OF ROUTINE PROCEDURE

Minor variations are necessary in order to put the various types
of purified uranium sample materials into solution prior to the basic
chemical separation. The preliminary steps appropriate to each of
the major types of uranium sample materials are given below.

### Uranyl Nitrate Solutions (10 to 100 μg Th/ml)

1. Pipet a 100-ml portion of the sample solution into a 400-ml beaker.

   Note: The 100-ml sample is appropriate for solutions containing 100 g, or less, of uranium per liter. If the uranium content is greater than 100 g/liter, all of the volumes specified in the separation procedure (other than the sample volume) should be doubled.

2. Continue with "Chemical Separation and X-ray Determination."

### Uranium Tetrafluoride (100 to 1000 ppm of thorium)

1. Transfer 10.0 g of the sample to a platinum dish and convert to $U_3O_8$ by pyrohydrolysis.)
2. Transfer the resulting $U_3O_8$ to a 400-ml beaker.
3. Add 10 ml of $HNO_3$ and heat until the $U_3O_8$ is dissolved.
4. Continue with "Chemical Separation and X-ray Determination."

### Uranium Metal, $UO_3$, or $U_3O_8$ (100 to 1000 ppm of thorium)

1. Transfer a 10.0-g portion of the sample material to a 400-ml beaker.
2. Add 10 ml of $HNO_3$ and heat until the sample is dissolved.
3. Continue with "Chemical Separation and X-ray Determination."

## CHEMICAL SEPARATION AND X-RAY DETERMINATION

1. Dilute the sample solution to 300 ml with distilled water.
2. Adjust the pH of the diluted solution to $1.0 \pm 0.1$ using $HNO_3$ or $NH_4OH$ as required.
3. Transfer the solution to a 500-ml separatory funnel. Rinse the beaker with 100 ml of $0.5M$ solution of TTA in xylene and add the rinsings to the separatory funnel.
4. Shake vigorously for 2 min. Allow the separatory funnel to stand for 2 min to allow the phases to separate.
5. Draw off the bottom (aqueous) phase and discard it to a (uranium) salvage container.
6. Add 200 ml of $0.2M$ $HNO_3$ to the separatory funnel, shake vigorously for 2 min, and allow to stand for 2 min.
7. Draw off the bottom (aqueous) phase and discard it to a salvage container.
8. Add 25 ml of $2.0M$ $HNO_3$ to the separatory funnel, shake for 2 min, and allow to stand for 2 min.
9. Draw off the bottom (aqueous) phase into a 100-ml beaker.

10.  Repeat steps 8 and 9, adding the bottom (aqueous) phase to the aqueous phase from step 9. Discard the organic phase.

11.  Pipet a 1.00-ml portion of the standard strontium nitrate solution (1.75 mg Sr/ml) into the beaker containing the combined aqueous phases from steps 9 and 10.

12.  Evaporate the solution to a volume slightly less than 10 ml by heating on a hot plate.

13.  Cool the solution and transfer to a 10-ml graduated cylinder. Rinse the beaker with small portions of distilled water, adding the rinsings to the graduated cylinder until a total volume of 10–11 ml is obtained.

14.  Transfer the solution to a polyethylene sample cell and secure a window of $1/2$-mil Mylar on the cell with a polyethylene band. Place the sample cell, window down, in a Norelco stainless steel sample holder (2-in.).

15.  Set up the instrument conditions as follows:
     X-ray tube voltage . . . . . . . . . . . . . . . . 60 kv
     X-ray tube current . . . . . . . . . . . . . . . 25 ma
     Analyzing crystal . . . . . . . . . . . . . . . . Lithium fluoride
     Optical path . . . . . . . . . . . . . . . . . . . . . Air
     X-ray detector. . . . . . . . . . . . . . . . . . . .Scintillator
     Electronic circuit panel. . . . . . . . . . . . . .Fixed count

16.  Record the scaling times (to the nearest 0.05 sec) for each of the goniometer settings, as indicated in Table I.

**TABLE I**
**Goniometer and Scaler Settings**

| Line | Goniometer setting, degrees $2\theta$ | Scaler setting, preset-fixed count |
|---|---|---|
| Sr bkg | 24.73 | 64,000 |
| Sr $K_\alpha$ | 25.16 | 256,000 |
| Th bkg | 27.08 | 64,000 |
| Th $L_{\alpha_1}$ | 27.46 | 256,000 |

## Calculations

1.  From the average scaling times, calculate the intensity ratio

$$IR[\text{Th}] = \frac{I[\text{Th}\,L_{\alpha_1}] - I[\text{Th bkg}]}{I[\text{Sr}\,K_\alpha] - I[\text{Sr bkg}]}$$

2. Convert the intensity ratio to thorium weight (micrograms) by referring to a current analytical curve or by the use of an equation derived from the curve.
3. Calculate the thorium concentration by referring the weight of thorium determined to the weight or volume of sample used

$$\text{ppm Th} = \frac{\mu\text{g Th determined}}{\text{sample weight (g)}}$$

or

$$\mu\text{g Th/ml} = \frac{\mu\text{g Th determined}}{\text{sample volume (ml)}}$$

## DISCUSSION

To obtain an estimate of the precision and accuracy of the X-ray method, a series of ten samples of $UO_3$ which contained a wide range of thorium concentrations was analyzed in duplicate, the duplicates being disguised and analyzed on separate days. The over-all limit of error (95% confidence level), computed from the differences between the duplicate determinations of the ten samples, was found to be $\pm 77$ $\mu$g Th in the range of 1000 to 10,000 $\mu$g Th, or $\pm 7.7$ ppm Th in the range of 100 to 1000 ppm.

Since uranium samples with established thorium values were not available, no data on the absolute accuracy are available. However, the X-ray determinations of thorium in the ten $UO_3$ samples are compared to averages of duplicate chemical determinations in Table II. An evaluation of the data of Table II reveals no statistical bias between the X-ray and chemical determinations.

## ESTIMATE OF TIME REQUIRED FOR ANALYSIS

It is estimated that one experienced operator can prepare and analyze eight samples of $UO_3$, $U_3O_8$, or U metal in an 8-hr day. One of the above analyses would require $1\,1/2$ hr. The time per sample required for uranyl nitrate samples is approximately 15 min less than that required for $UO_3$, $U_3O_8$, or U metal. Because $UF_4$ samples must first be converted to $U_3O_8$, approximately 9 hours are required for the preparation and analysis of eight $UF_4$ samples.

TABLE II
Comparison of Chemical and X-ray Determination of Thorium

| | X-ray, ppm | | | Chemical, ppm Avg.* | Difference, X-ray—Chemical |
|---|---|---|---|---|---|
| | No. 1 | No. 2 | Avg. | | |
| 3P-794 | 370 | 363 | 366.5 | 362.5 | + 4.0 |
| 796 | 387 | 391 | 389.0 | 387.0† | + 2.0 |
| 798 | 389 | 388 | 388.5 | 393.0 | – 4.5 |
| 801 | 571 | 573 | 572.0 | 556.0 | +16.0 |
| 802 | 665 | 656 | 660.5 | 665.0 | – 4.5 |
| 803 | 655 | 646 | 650.5 | 653.5 | – 3.0 |
| 806 | 314 | 314 | 314.0 | 316.5 | – 2.5 |
| 808 | 249 | 246 | 247.5 | 251.5 | – 4.0 |
| 810 | 237 | 242 | 239.5 | 240.0 | – 0.5 |
| 811 | 172 | 175 | 173.5 | 179.5 | – 6.0 |

*Average of duplicate chemical determinations (95% confidence level limit of error for averaged duplicates = ±17 ppm Th).
†Single chemical determination (limit of error, 95% confidence level = ±24 ppm).

# ACKNOWLEDGMENTS

The authors wish to acknowledge the assistance of Mr. C. E. Pepper, who provided constructive criticism throughout the various stages of the development work. Thanks are also due Mr. W. D. Kelley, who carried out the radioactive tracer studies which were of great assistance in developing and evaluating the extraction procedure.

# REFERENCES

1. M.O. Fulda, *Determination of Traces of Thorium in Uranium Solutions*,USAEC Report DP-165, June 1956.
2. P. G. Laux and E. A. Brown, "Determination of Thorium in Uranium Metal," *Summary Tech. Rpt.*, USAEC Report NLCO-760, pp. 111-115, Sept. 1958.
3. P. G. Laux and E. A. Brown, *Determination of Thorium in Uranium Ores and Feeds by Solvent Extraction Employing Thenoyltrifluoroacetone*, USAEC Report NLCO-742, May 23, 1958.
4. J. F. Moskal and H. A. Heller, "X-ray Spectrometric Determination of Thorium in Uranium Concentrates," *Summary Tech. Rpt.*, USAEC Report NLCO-785, pp. 143-148, Apr. 15, 1959 (Classified).
5. G. R. Blank and H. A. Heller, "X-ray Spectrometric Determination of Copper in Tin," *Summary Tech. Rpt.*, USAEC Report NLCO-805, pp. 81-85, Apr. 29, 1960.
6. G. R. Blank and H. A. Heller, "X-ray Spectrometric Determination of Copper, Tin and Uranium in Bronze Heat-Treating Material," in *Advances in X-ray Analysis*, Vol. 4, edited by W. M. Mueller, Plenum Press, New York, 1960.
7. G. Pish and A. A. Huffman, "Quantitative Determination of Thorium and Uranium in Solutions by Fluorescent X-ray Spectrometry," *Anal. Chem.* 27, 1875, Dec. 1955.
8. R. C. Sackville, "Colorimetric Determination of Thorium in Uranium Bearing Solutions," *Eldorado Analytical Methods Manual AM-117*, Eldorado Mining and Refining Co. Limited, May 1955.

# X-ray Spectrographic Analysis
# of Antimonials

## R. Bruce Scott

Research Laboratories
Parke, Davis & Co.
Ann Arbor, Michigan

---

The accurate analysis of organic (or inorganic) antimonials requires careful control of several important factors. Instrument variables, especially the intensity of the primary radiation, are critical and must be monitored by use of a reliable standard. Tartar emetic, which contains 36.5% antimony and is readily soluble in water, is a convenient standard.

If one works in solution, which is unquestionably the more accurate method, there may be difficulties because of low solubility. Several solvents have been used with success; dimethylformamide, a 5× dilution of hydrochloric acid, and 1N sodium hydroxide have been the most useful. In some cases decomposition of the compound in a strong mineral acid with subsequent dilution in water is necessary. Whatever solvent is used, it is important that the standard solution have the same composition, unless an internal standard can be used.

In the interest of conservation of limited sample quantities, small volumes of dilute solutions are used. Generally, the antimony concentration runs in the range of 1 to 2 mg/ml. One type of cell requires 2 ml of solution; others take up to 14 ml. The most frequently used cell holds 5 ml and has a depth of 7.5 mm.

The use of dry dilutions in lactose, with subsequent briquetting, has been investigated and can be useful where there is a real solubility problem. However, because of the increased time of sample preparation, the more cumbersome procedure for making dilutions, and the uncertainties, such as particle size, introduced by the physical state of the mixture, this is not the method of choice.

The analysis of organic antimonials poses no new and difficult problems for the X-ray spectroscopist. Recently, when an accurate assay was desired for a series of these compounds, several modes of operation were evaluated, and a rapid analysis was devised.

The preferred method begins with dissolution of a few milllgrams of the sample in a suitable solvent. If the compound is not amenable to this treatment, briquetting, either with or without dilution in a powder such as lactose, is necessary. The solution method is preferred because of its elimination of particle-size effects and its insurance of homogeneity. Further, the preparation of a series of standard solutions is relatively simple.

At an early stage of the study it was necessary to choose between the use of $K$ or $L$ fluorescent radiation. The Sb $K_\alpha$ line with a wavelength of 0.470 A is reflected from a LiF analyzing crystal at a $2\theta$ angle of 13.5°. The Sb $L_{\alpha_1}$ line with a wavelength of 3.439 A is reflected from a NaCl crystal at 75.16° $2\theta$. With the former, an air path and scintillation counter is a convenient arrangement; for the $L$ line, a helium atmosphere in combination with a flow-proportional counter is necessary. The intensities, relative to background, for the two systems are shown in Table I. The choice here, at least at this

### TABLE I
**Sample: 0.6-g Briquet of Lactose and Tartar Emetic,
Containing 1% Antimony
Source: Tungsten OEG-50, Operated at 40 kv, 20 ma**

| Sb $K_\alpha$ radiation, LiF crystal, air path, scintillation counter | | Sb $L_\alpha$ radiation, NaCl crystal, helium path, flow-proportional counter | |
|---|---|---|---|
| Peak: | 4375 counts/sec | Peak: | 156 counts/sec |
| Background: | 2240 counts/sec | Background: | 8 counts/sec |
| Ratio: | 1.95 | Ratio: | 19.5 |

concentration of antimony, lies between a method giving high intensities with relatively low peak-to-background ratios, and a method giving low intensities with much higher peak-to-background ratios. The decision to utilize the $K$ radiation was influenced by the limited solubilities and quantities of some of the samples. This of course would have the effect of keeping $L_\alpha$ fluorescence at very low levels.

Aside from controlling the actual amount of sample in the exciting X-ray beam, the most critical need of the determination is control of the beam intensity. The steep slope of the applied poten-

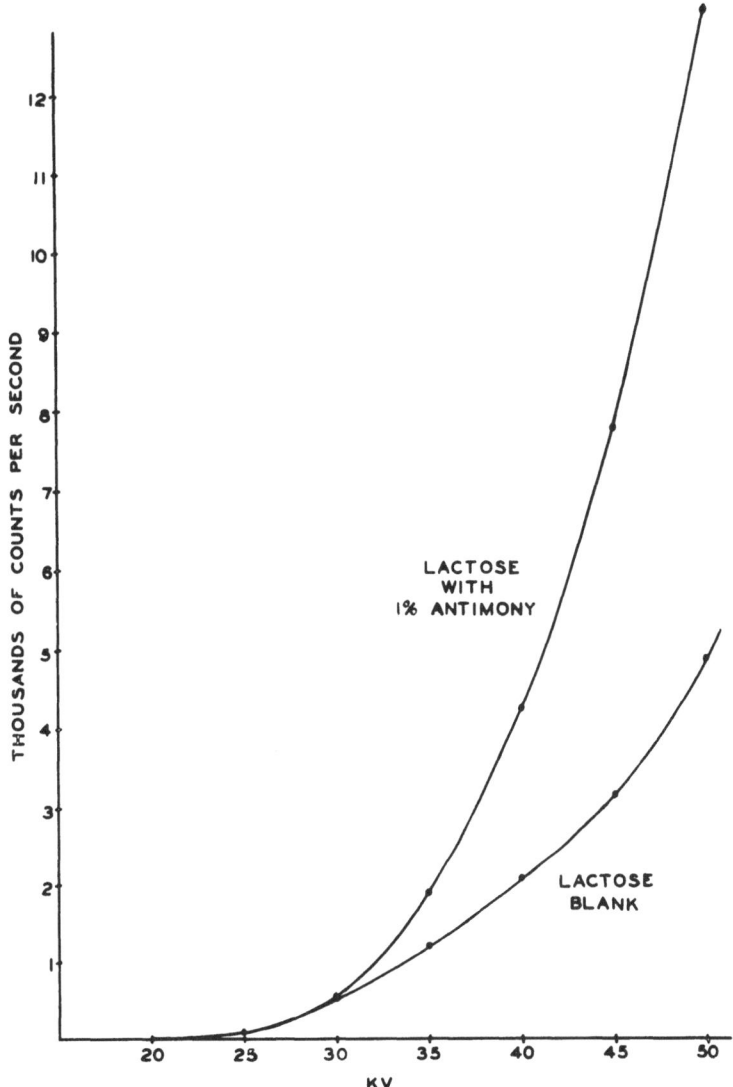

Fig. 1. Intensity *vs* voltage on source tube (20 ma, 13.5° 2θ).

tial *vs* fluorescent intensity curve (Fig. 1) makes this plainly evident. Because a variation of 0.1 kv produces a difference of nearly 60 counts/sec in the fluorescent intensity of a 1% antimony sample, and because voltage control allows this much variation, a method of monitoring beam intensity is highly desirable. Internal standards are commonly added for this purpose, as well as to compensate for matrix absorption effects. However, the addition of other compounds began to create solution difficulties, and therefore, an alternate method was used.

In this method, scattered radiation of an appropriate wavelength is measured immediately before or after reading the Sb $K_\alpha$ peak

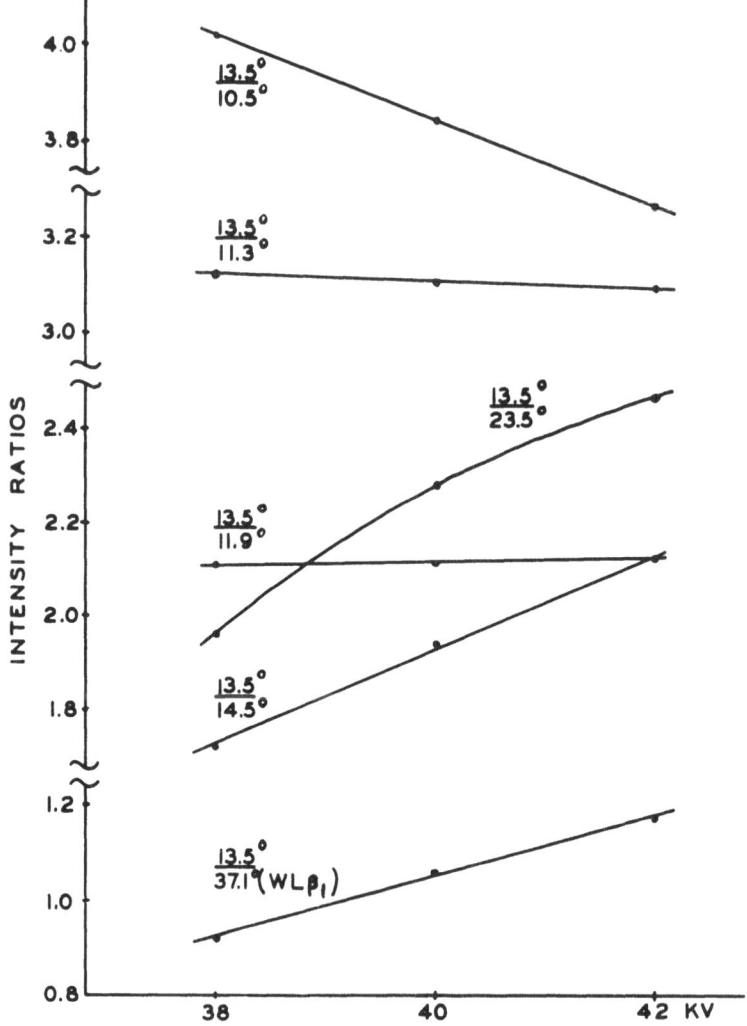

Fig. 2. Peak-to-background ratio *vs* tube voltage (1% antimony in lactose).

intensity. While this does not give 100% correction, precision is increased by a factor of two over peak intensity readings taken alone. For example, in a series of 20 readings on one specimen, taken over a period of 2 days, the relative standard deviation of peak-to-background ratios was 0.5%. The corresponding figure for the peak intensities on the same series of readings was 1%. It should be kept in mind that these figures include the effects of all instrument variables, including the standard counting error, which for the 204,800 pulses counted is 0.22%. The use of scattered X-rays as internal standards has been described more fully by G. Andermann and J. W. Kemp.* As they commented, the selection of a suitable wavelength at which to measure the scattered radiation is largely empirical. The choice may be different depending on whether one is seeking primarily to compensate for variations in source radiation or for differences in matrix composition. Figure 2 illustrates the effect of choosing different scattering wavelengths on the intensity ratio *vs* applied potential curve. It should be noted that the incorporation of $K_\beta$ radiation in the measured background, as happens at 11.9° $2\theta$, will give a bent working curve (lower curve, Fig. 3). This is expected on a theoretical basis.

When the effects of variations in source intensity have been minimized, the principal factor requiring control is the composition of the solution. In this work the compounds were primarily organic in nature, with the antimony content being in the range 20 to 40%. The heaviest element other than antimony was chlorine, which occurred as a minor constituent in some of the compounds. The solutions were made up to have an antimony concentration of 1 to 2 mg/ml. Thus, the nature of the absorbing medium was essentially that of the solvent. Most of the work was done with four solvents: water, water and dimethylformamide (50:50 v/v), 1$N$ HCl, and 1$N$ NaOH. The unknowns were generally dissolved in concentrated DMF, HCl, or NaOH, which was then diluted to the proper strength. Tartar emetic, containing 36.5% antimony, is a convenient reference compound although it is, unfortunately, not soluble in straight DMF. Table II, showing the effect of various solvents on Sb $K_\alpha$ intensity, emphasizes the importance of using the same solvent for standard and unknown.

*Analytical Chemistry* **30**, 1306–1309 (1958).

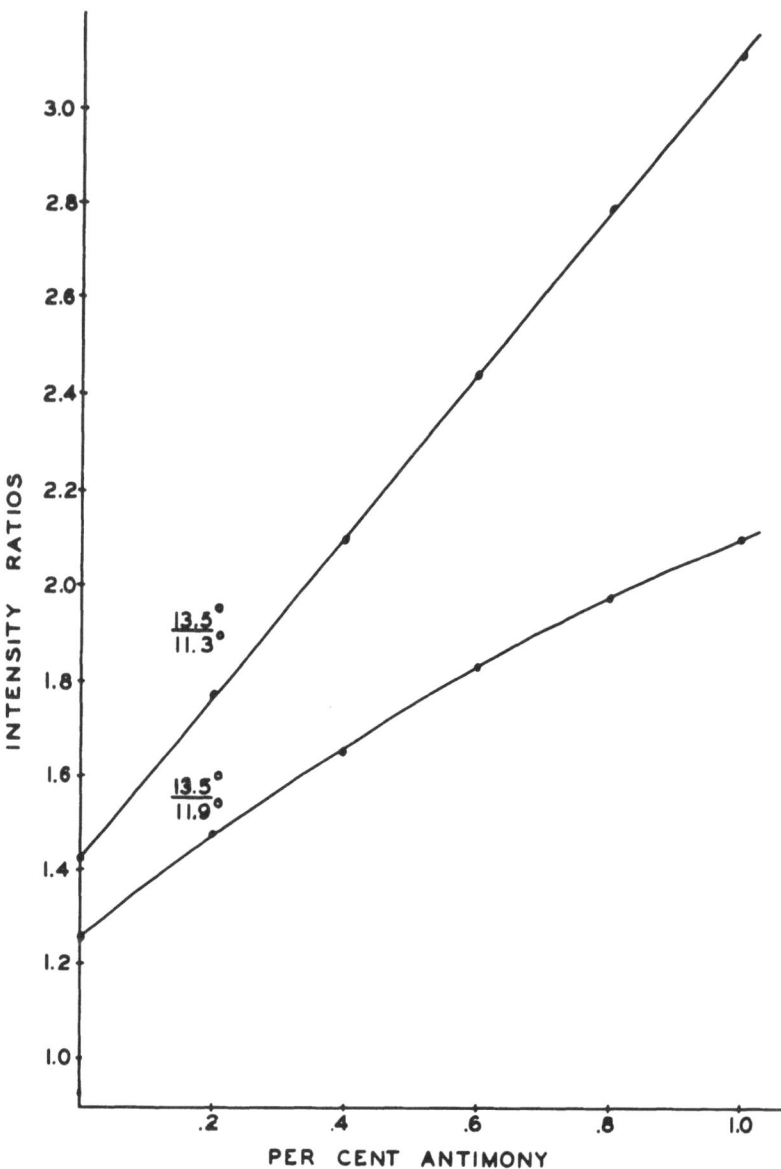

Fig. 3. Tartar emetic in lactose (0.6-g briquet, 40 kv, 20 ma).

In order to handle small volumes, specially made cells of 2 and 5 ml capacity have been fabricated of aluminum and Plexiglas. The 2-ml cell has a sample depth of 3 mm; the 5-ml cell, a sample depth

## TABLE II
### Effect of Solvents on Sb$K_\alpha$ Intensity

| Solvent | Intensity, counts/sec |
| --- | --- |
| $H_2O$ | 4070 |
| 50:50 $H_2O$—DMF (v/v) | 4135 |
| 1N HCl | 3830 |
| 1N NaOH | 4075 |
| $H_2O$ blank | 3000 |

Conditions: 40 kv; 20 ma; LiF crystal; antimony concentration, 2 mg/ml.

of 7.5 mm. These fit snugly into the sample tray, which slides in horizontally under the source tube. Where the nature of the solvent permits, the aluminum cell is used because less primary radiation is scattered from it.

The assay just described is an example of the adaptability of X-ray fluorescence techniques to varied analytical problems. In this case, the slowest part of the assay was the time spent in finding suitable solvents for the various compounds. That is one reason a program to explore means of facilitating more rapid and accurate analyses of dry powders has been initiated in this laboratory.

# Potassium Pyrosulfate Fusions for X-ray Spectroscopy

## Thomas J. Cullen

United States Metals Refining Co.
Carteret, New Jersey

An X-ray spectrographic method is reported for the analysis of brass and bronze samples by the potassium pyrosulfate fusion technique. The fusions are carried out at moderate temperatures in glass beakers. Absorption and enhancement effects are removed. Standards prepared from pure metals are used to determine copper, lead, nickel, tin, and zinc in the samples. The fusions are ground and briquetted so that a uniform surface is exposed to the X-rays.

## ANALYSIS OF COPPER-BASED ALLOYS

Deviations from proportionality complicate X-ray spectrographic methods of analysis. The main causes of these deviations fall into three classes: (1) absorption and enhancement effects; (2) heterogeneity in the samples, including surface effects; and (3) instrumental instability.

Deviations due to classes 1 and 2 can be greatly reduced by fusion in a properly chosen flux. Fusions will dilute, homogenize, and react with the sample. Dilution alone will minimize the matrix change between individual samples and, thus, reduce absorption and enhancement effects between individual particles of the sample; however, dilution will not remove absorption and enhancement effects within individual particles. Dilution may also introduce a heterogeneity error. Since in mechanical dilution no reaction between the diluent and sample occurs, the possibility of an error due to differing valence states of an element exists.

Fusion or dilution in a lightweight matrix such as carbon, borax, silica, paper pulp, *etc.*, will reduce the loss of fluorescent intensity. With lightweight-matrix dilutions, one finds that large dilutions are necessary to reduce or eliminate matrix absorption and enhancement effects; incoherent Compton scattering of the intensities also occurs. Absorption and enhancement effects occur due to the penetration of the primary X-ray beam, which excites elements below the surface of the sample, thus causing increased interelemental reactions of the fluorescent energies. Claisse [1], in his paper on the borax-bead technique, suggests the addition of an absorbent compound, such as potassium pyrosulfate or barium peroxide, to the flux to reduce this effect.

Deviations due to heterogeneous samples are eliminated by fusion techniques since the sample is reduced to a homogeneous solid solution, assuming total reaction of the sample. It must be pointed out that since fusion techniques use small sample weights, the sample must be well prepared to the extent that the small sample weight fused will be representative of the whole sample.

Potassium pyrosulfate [2] has been found to be a very useful flux in the fusion technique of preparing samples by X-ray spectroscopy. It has been found that 200 mg of the sample fused in 10.0 g of potassium pyrosulfate is adequate to remove severe matrix effects. This is due to its high reactivity in dissolving samples, and to the absorbing characteristics of the potassium, and, to some extent, of the sulfur present in the matrix. The inhibition of the absorption of copper fluorescent radiations by iron in samples of mattes and slags has been reported. These samples contained from 0 to 60% copper and from 0 to 20% iron. The inhibition of the absorption effect was so pronounced that synthetic standards containing only copper could be used as calibration standards. Deviations from electrolytic copper analysis were in the range of 0.1% absolute. These results were obtained by taking a minimum of 100,000 counts and averaging the results of three separate sample portions.

The following data were obtained using a General Electric XRD-5 X-ray spectrometer, a scintillation counter, a Victoreen linear amplifier with a 0- to 20-v window and 0- to 100-v baseline pulse-height selector, a 10-ml Soller slit, a lithium fluoride analyzing crystal, and a tungsten-target X-ray tube operated at 50 kv and 50 ma. The magnitude of the number of counts recorded for the intensities measured was such that the degree of variation was less than the significant numbers reported in the data.

A brief description of the fusion process is as follows: 5.000 g of potassium pyrosulfate is weighed into a fused silica crucible and

100 mg of sample is mixed well with the potassium pyrosulfate to insure rapid and uniform solution. The fusion mixture is then heated to a moderate temperature, using a bunsen burner, until the mixture melts and the sample dissolves. When the fusion is complete, it is allowed to cool and is ground either by hand or mechanically. The fusion is then briquetted at a pressure in excess of 12,000 psi to form a briquet 1.25 in diameter. The sample is then ready to be used in the X-ray spectrometer

Potassium pyrosulfate decomposes with the evolution of sulfur trioxide at temperatures in excess of 700°C. Since the flux will melt at 300°C, fusions can be carried out between 300 and 700°C without significant losses of weight. If the sample contains metallics or refractories, or for any reason requires prolonged heating at the higher temperature range, a weight-loss correction should be applied to obtain the best results. If the standards and the samples require the same time and temperature to dissolve, the weight loss will be the same and no correction will be necessary. Swirling the melt while fusing will reduce localized heating, homogenize the melt, and reduce fusion time. The time requirement for fusion of 100 mg of sample in 5.0 g of potassium pyrosulfate is between 1 and 3 min.

The fusion willl usually crack into several pieces upon cooling, thus facilitating removal from the beaker or crucible. The grinding of the pieces of the fusion is easier and more rapid if it has not cooled to room temperature, but is still hot to the touch. The fusions have been ground in our laboratory in a mixer mill, using disposable plastic mixing balls and vials. The grinding operation takes 10 to 15 min, and four fusions are ground at one time.

Since fused potassium pyrosulfate will absorb moisture, one might expect an error to be introduced. It has been determined that in the grinding process the fine particles of powder absorb moisture to the point of equilibrium; thus, the amount of moisture absorbed is constant and reproducible before the fusion is briquetted. However, if the fusion is to be kept for a prolonged period of time, it is recommended that it be stored in a desiccator. The instability of the briquets appear to arise from handling rather than from other factors. The briquets can be reground and briquetted when this handling damage occurs.

Since potassium pyrosulfate has the ability to dissolve metallic samples, it was thought that the precision and accuracy of the technique could be demonstrated by analyzing a series of NBS

**TABLE I**

**Analyses of NBS Copper-Based Alloys**

| NBS No. and Type | Cu, % | | Zn % | | Sn, % | | Pb, % | | Ni, % | |
|---|---|---|---|---|---|---|---|---|---|---|
| | NBS | X-ray | NBS | X-ray | NBS | X-ray | NBS | X-ray | NBS | X-ray |
| 37a Sheet brass | 70.78 | 70.90 | 26.65 | 26.70 | 0.97 | 0.93 | 0.94 | 0.90 | 0.58 | 0.60 |
| 52b Cast bronze | 88.25 | 88.30 | 2.96 | 2.90 | 8.00 | 7.95 | 0.011 | 0.00 | 0.72 | 0.75 |
| 62b Manganese bronze | 57.39 | 57.35 | 37.97 | 38.15 | 0.96 | 0.96 | 0.28 | 0.27 | 0.27 | 0.27 |
| 63b Phosphor bronze | 77.94 | 77.87 | 0.71 | 0.65 | 9.78 | 9.88 | 9.36 | 9.31 | 0.33 | 0.30 |
| 157 Nickel silver | 72.14 | 72.28 | 9.69 | 9.75 | – | – | 0.023 | 0.00 | 17.90 | 17.95 |
| 124b Ounce metal | 86.69 | 86.60 | 5.40 | 5.32 | 4.93 | 4.99 | 4.64 | 4.69 | 0.76 | 0.79 |
| 124d Ounce metal | 83.60 | 83.75 | 5.06 | 5.05 | 4.56 | 4.55 | 5.20 | 5.26 | 0.99 | 1.02 |
| 162 Monel | 28.93 | 28.99 | – | – | – | – | – | – | 66.38 | 66.50 |
| 113 Zinc ore | – | – | 61.1 | 61.2 | – | – | – | – | – | – |

copper-based alloys, using synthetic standards for calibration. The standards were prepared by weighing into 5 g of potassium pyrosulfate various amounts of pure metals plus pure copper, to a total weight of synthetic sample of 100 mg. Standards containing various amounts of pure copper were used in the determination of copper, and the sample weight was mathematically corrected to 100 mg. Copper, lead, nickel, tin, and zinc standards were prepared so that the contents covered the range of the amounts present in the NBS standards. Table I shows the results of this test. The results reported are the averages of triplicate determinations.

A NBS standard zinc ore, No. 113, was used to determine the precision and accuracy of the zinc determination. Standards containing 50 and 70% zinc were prepared from pure zinc oxide. The two standards and the sample were prepared and run on ten different days over a period of five months. The results are reported in Table II.

The determination of elements existing in a form insoluble in potassium pyrosulfate is not as accurate or precise as the determination of elements in soluble form. An example of this is germanium dioxide. The net intensity from a series of briquetted fusions of 100 mg of 100-mesh germanium dioxide had a deviation of 7% from the mean intensity. This deviation was decreased to 3% by fusing 300-mesh germanium; thus, the ground fusion was actually a

## TABLE II
### Determination of Zinc in NBS 113
### (Zinc Ore - 61.1% Zn)

| Zn found, % | Difference |
|---|---|
| 61.20 | 0.10 |
| 61.15 | 0.05 |
| 61.31 | 0.21 |
| 61.10 | 0.00 |
| 61.14 | 0.04 |
| 61.40 | 0.30 |
| 61.12 | 0.02 |
| 61.12 | 0.02 |
| 61.16 | 0.06 |
| 61.22 | 0.12 |
| $\sigma = 0.096$ | |

mechanical dilution of the oxide. The only advantage in the fusion of samples containing germanium dioxide is the removal of the effect of other elements present in the sample.

## REFERENCES

1. F. Claisse, *Norelco Reptr.* 4, 3, 1957.
2. T. J. Cullen, *Anal. Chem.* **32**, 516, 1960.

# Fluorescent X-ray Spectrography: Determination of Trace Elements

## William J. Campbell and John W. Thatcher

United States Department of the Interior
Bureau of Mines, College Park Metallurgy Research Center
College Park, Maryland

Serious consideration should be given to the merits of using the available fluorescent X-ray spectrographic instrumentation and techniques for determining trace elements. Limits of detectability for trace elements in metal, oxide, solution, or mineral samples range from 0.1 to 100 ppm, depending on the element being determined, over-all sample composition, and complexity of the X-ray spectra; limits of detectability range from 0.01 to 1 $\mu$g for elements that have been preconcentrated by a chemical or physical process, for example, by ion-exchange membrane.

Fluorescent X-ray spectrography can be used to obtain absolute values (by preparation of standard samples) or relative values based on standards analyzed by other means. Since X-ray techniques are nondestructive, both standards and unknowns are conserved for use in other evaluation procedures.

## INTRODUCTION

There have been many significant developments in the instrumentation and analytical procedures used with fluorescent X-ray spectrography during the past decade. Fluorescent X-ray spectrography is now widely accepted by both research and control laboratories for determining major and minor constituents [10]. The purpose of this report is to summarize the present status of the technique in studies involving trace elements. The findings are restricted to limits of detectability that can be obtained with conventional fluorescent X-ray spectrographic instrumentation and techniques.

Because the meaning of the term *trace* is ambiguous, some comments are necessary on its definition with respect to this report.

Originally trace, in a chemical sense, meant a quantity too small to be measured or not worth determining. Today, with the development of high-purity materials, the term signifies a minute amount which can be determined, usually with some degree of accuracy required. However, the percentage of a constituent required for classification in the trace-element range is arbitrary. Hillebrand [34] said, with reference to rock analysis, "It may be said with regard to the use of the word *trace* that the amount of a constituent thus indicated is supposed to be below the limit of quantitative determination in the amount taken for analysis. It should in general, for analysis laying claim to completeness and accuracy, be supposed to indicate less than 0.02 or 0.01%." However, Sandell [51] states, "There is no reason for making this boundary a rigid one. It is sometimes convenient to consider as a trace constituent one that occurs to the extent of a few hundredths of a percent. Thus in silicate rocks, copper usually falls in the range 0.001 to 0.05%, and it is permissible to speak of copper as a trace constituent in this class of material." In this report, the detection of quantities less than 0.05% (500 ppm) is emphasized.

## SAMPLE CLASSIFICATION

Liebhafsky, Pfeiffer, Winslow, and Zemany [44] suggest that trace analysis can be conveniently subdivided according to sample type: (1) traces as minor constituents in samples not unusually small and (2) traces as major constituents of a minute sample. In many instances, for example, chemical preconcentration, it would be necessary to isolate the desired elements from the bulk sample; this amounts to conversion of a class 1 to class 2 sample.

In this report, class 1 samples are divided into metals, oxides, solutions, ores, and miscellaneous materials. For convenience and uniformity, limits of detectability for class 1 samples are given in parts per million.

Class 2 samples can be subdivided either according to the analytical procedure used to concentrate the desired elements, *e.g.*, ion-exchange, solvent extraction, *etc.*, or, as in this report, by the manner in which the sample is presented for X-ray analysis, *e.g.*, on an ion-exchange membrane or filter paper support. Limits of detectability for class 2 samples are generally given in terms of micrograms, since the sensitivity expressed in parts per million depends on both the limit of detectability in micrograms and the original weight of the sample before preconcentration of the trace elements.

## FUNDAMENTALS

### Line Intensity

The ultimate objective in trace analysis by fluorescent X-ray spectrography is the quantitative measurement of a weak spectral line above background, where the background may contribute a significant number of the X-ray quanta counted. For maximum sensitivity, the X-ray tube should be operated at its highest voltage and current ratings. With 60-kv power supplies, the $K$ series X-ray lines are used for elements below atomic number 40, the $L$ lines for elements above atomic number 60, and either the $K$ or $L$ lines for elements whose atomic numbers range from 40 to 60 [11]. If 100-kv power supplies are available, the $K$ series lines may be effectively excited for elements greater than atomic number 60; however, crystals with both high reflectivity and suitable $d$-spacings for these $K$ lines are not known.

Applications to trace analysis require the strongest possible signal; therefore, the resolution should be limited to only what is actually required to separate the lines of interest. High resolution X-ray optics can be achieved only by fine collimation, thereby losing intensity. Flat-crystal X-ray optics are covered in detail by Spielberg, Parrish, and Lowitzsch [54], and, in a previous paper, by Campbell, Leon, and Thatcher [15]. Focusing (curved-crystal) and nonfocusing (flat-crystal) optics are compared in books on the subject of X-ray spectrography [4,44].

For a given set of operating conditions, the strength of the signal is a function of the concentration of the element being determined and the over-all sample composition. If the source of excitation is assumed to be monochromatic and nondivergent and no excitation results from enhancement in the sample, then the following equation expresses the relationship of intensity to both the sample composition and the concentration of the element being determined for a class 1 sample of effective infinite thickness:

$$_{\lambda_2}I_a = \frac{KW_a}{_{\lambda_1}(\mu/\rho)_s + _{\lambda_2}(\mu/\rho)_s} \tag{1}$$

where $I_a$ is the intensity of characteristic spectral line of element $a$, $W_a$ is the weight-percent of element $a$, and $(\mu/\rho)_s$ is the mass absorption coefficient of sample $s$ for incident radiation $\lambda_1$ and secondary radiation $\lambda_2$.

Examination of equation (1) reveals that for a given element the strength of a particular characteristic line is inversely proportional to the summation of $_{\lambda_1}(\mu/\rho)_s$ and $_{\lambda_2}(\mu/\rho)_s$. The sensitivity for trace elements will therefore be greater (a stronger signal) in samples of low X-ray absorption characteristics. To illustrate this important point, intensity values were calculated for Fe $K_\alpha$, Mo $K_\alpha$, and Sn $K_\alpha$ in various matrices relative to those calculated for a nickel base. These calculations (Table I) indicate that the $K$ spectral lines of iron, molybdenum, and tin should be over 50 times stronger in beryllium metal than in nickel. Conversely, spectral intensities from impurities in high-atomic-number metals such as tungsten may be reduced by a factor of 3.

## Background

The background intensity is also an important factor in determining the limits of detectability. Background is the summation of

### TABLE I
### Calculated Line Intensities as a Function of Sample Composition*

| Sample | Intensity, relative values† | | |
|--------|---------|---------|---------|
|        | Fe $K_\alpha$ | Mo $K_\alpha$ | Sn $K_\alpha$ |
| Be | 85 | 126 | 71 |
| O | 12 | 38 | 32 |
| Mg | 3.4 | 11 | 12 |
| S | 1.5 | 4.9 | 5.1 |
| Ca | 0.85 | 2.6 | 2.7 |
| Ni | 1.0 | 1.0 | 1.0 |
| Zn | 2.3 | 0.83 | 0.85 |
| Sr | — | 0.45 | 0.44 |
| Mo | — | — | 0.35 |
| Ag | 0.63 | 1.7 | 0.55 |
| W | 0.86 | 0.46 | 0.47 |
| Au | 0.66 | 0.39 | 0.37 |
| Pb | 0.60 | 0.35 | 0.33 |

*$\lambda_1$ for excitation of iron, molybdenum, and tin assumed to be 1.43, 0.497, and 0.417 A, respectively see equation (1).
† Based on intensities from nickel equal to 1; secondary excitation by sample not considered; intensities cannot be compared between columns, for example Fe $K_\alpha$ to Mo $K_\alpha$.

various components: electronic noise and cosmic radiation, spectral impurities originating in the X-ray-tube target and window, Compton or modified scattering, characteristic lines from spectrographic components, and primary unmodified scattering from the sample and its surroundings.

Flow-proportional and scintillation counters have a noise level of less than 1 count/sec when used with a pulse-height discriminator and linear amplifier. Geiger counters of the type used in X-ray spectrography give background counts of 1 to 2 counts/sec. Lead shielding of the detectors to reduce cosmic radiation is not warranted for these applications.

The purity of the X-ray-tube spectra is very important in certain trace analysis procedures. The Machlett tubes,* which are widely used as the primary X-ray source, were designed for radiological applications in which spectral impurities are not important. Principal impurity spectral lines are those of copper, nickel, and iron, plus the anode; the anode is either tungsten or molybdenum. Higher-purity tubes now available, either Machlett OEG-60 or Philips FA-60, have low iron, nickel, and copper intensities [40].

When the primary X-ray beam is scattered by the sample, X-rays lose energy (Compton effect) when shifting to a longer wavelength. This Compton scattering, which is more pronounced for samples comprised of low-atomic-number elements, results in broader or split primary spectral lines and thus increases the possibility of interference with analytical spectral lines. The wavelength $\lambda$ of the modified radiation can be calculated from the following expression:

$$\lambda - \lambda_0 = (h/mc)\,(1 - \cos\phi) = 0.02427\,(1 - \cos\phi)\,\text{A} \qquad (2)$$

where $\phi$ is the angle between the primary and scattered beam. In most commercial X-ray spectrographs, $\phi$ approximates 90°; therefore, $\Delta\lambda$ is close to 0.02427 A. Dwiggins [22] determined the ratio of carbon to hydrogen in petroleum hydrocarbons by measuring the modified and unmodified scattering of a strong tungsten $L$ line. This scattering ratio is dependent on the relative abundance of elements of very low atomic number. The applications of this technique may be broadened to include determination of the ratio of metal to oxygen, nitrogen, or carbon.

*Reference to specific brands is made to facilitate understanding and does not imply endorsement of such brands by the Bureau of Mines.

Other spectrographic components that contribute to the background are the collimator plates, analyzing crystal, and detector. Secondary spectral lines originating in the primary collimator (between the sample and analyzing crystal) must undergo Bragg diffraction before reaching the detector. This radiation is, therefore, angular-dependent and will appear as a broadened line or band. Radiation from the secondary collimator is also angular-dependent; its intensity will be additive to the strong spectral line from the sample. As an illustration, consider a sample high in zinc content with a secondary collimator composed of nickel plates. The nickel is strongly excited by Zn $K_\alpha$ and $K_\beta$ radiation; however, the zinc radiation is only diffracted over a narrow angular range. The resultant nickel intensity will overlap the Zn $K$ lines and add to the over-all intensity at the Zn $K_\alpha$ and $K_\beta$ peak position. Nickel $K_\alpha$ and $K_\beta$ lines originating in the primary collimator will be found at different angles than the zinc radiation, but will be broader as the effective collimator length is less for the Ni $K$ radiation.

Analyzing crystals composed of elements above atomic number 12 are possible sources of significant secondary radiation. This secondary radiation is not angular-dependent and results in a general increase in background. In most instances the radiation originating in the crystal can be eliminated by pulse-height discrimination; however, the analyst should consider the possibility of its occurrence. For example, the Cl $K_\alpha$ radiation from a sodium chloride analyzing crystal contributed a significant portion of the background with high-calcium samples [17].

Proportional and scintillation detectors also have components which can be excited by higher energy radiation, all of which is angular-dependent. A good example of this type of background radiation occurs in the determination of hafnium in zirconium, where the separation of Hf $L_\alpha$ first order from the strong Zr $K_\alpha$ second order line is a problem. With hafnium-free samples there is still a measurable intensity above the background value at the Hf $L_\alpha$ position, even though the pulses characteristic of Zr $K_\alpha$ lines and their escape peaks are discriminated against by the pulse-height analyzer. A probable source of this unexpected intensity may be either the tungsten anode of the proportional counter, or copper and zinc in the brass housing. All three spectral series, Cu $K_\alpha$, Zn $K_\alpha$, and W $L_\alpha$, have energies approximating that of Hf $L_\alpha$ and, therefore, would not be completely resolved by pulse-height discrimination. Analyzing crystals of silicon or germanium are now available that have a very low reflectivity for even-ordered reflections; however, their first-order reflectivity is less than that of lithium fluoride, and the angular dispersion is lower because of their higher $d$-spacing. For special applications, $e.g.$, hafnium in zirconium, construction of special proportional counters, using materials that do not emit interfering spectral lines, may be warranted.

Most of the background radiation from class 1 samples is unmodified scattered primary radiation. Other studies by the Bureau of Mines [15] showed pulse-height discrimination (PHD) to be very effective in the long-wavelength region (greater than 2 A) since the scattered radiation is comprised principally of multiordered high-energy X-rays. In addition, the flow-proportional counter has a low counting efficiency for high-energy radiation, thus providing additional discrimination. Pulse-height discrimination is less effective in the short-wavelength region as the scattered radiation is principally first order. Other studies by the Bureau of Mines (Table II) show that pulse-height discrimination will not improve the line-to-background ratio by a factor of 2 until the X-ray-tube voltage exceeds the critical voltage of that element by a factor of 3. It was also concluded that the scintillation and flow-proportional counters

## TABLE II
### Reduction in Background by Pulse-Height Discrimination

| Detector | Goniometer setting* | Critical voltage, kev† | Background reduction, relative values‡ |
|---|---|---|---|
| Flow-proportional§ | S $K_\alpha$ | 2.5 | 11.5 |
| " | K $K_\alpha$ | 3.6 | 7.2 |
| " | Ti $K_\alpha$ | 5.0 | 2.6 |
| Scintillation ‖ | Ti $K_\alpha$ | 5.0 | 35.1 |
| " | Mn $K_\alpha$ | 6.5 | 18.2 |
| " | Ni $K_\alpha$ | 8.3 | 10.4 |
| " | Zn $K_\alpha$ | 9.7 | 4.1 |
| " | Se $K_\alpha$ | 12.7 | 2.4 |
| " | Mo $K_\alpha$ | 20.0 | 1.4 |
| " | Ag $K_\alpha$ | 25.5 | 1.1 |

*Measurements made on water with goniometer at the specified position, 55 kv on X-ray tube.
†Voltage required to ionize elements listed in second column.
‡Intensity ratio: analyzer window at infinity/analyzer window of 7v.
§Quartz crystal $2d$ = 6.68 A.
‖ LiF crystal $2d$ = 4.03 A.

effectively cover the normal wavelength range found in fluorescent X-ray spectrography. There is no improvement in the line-to-background ratio for the xenon-filled proportional counter as compared with the scintillation counter because the pulse-height discrimination is against multiordered radiation instead of spectral lines from elements several atomic numbers apart.

Air scattering of primary X-rays can contribute significantly to the background with class 2 samples. Barstad and Refsdal [3] reported that background intensities were greatly reduced for very thin samples, less than $1\mu$ thick, when the sample chamber is evacuated. As the sample thickness is increased, the X-ray scattering by the sample increases; the contribution to background from air scattering is not significant for class 1 samples.

## Matrix Correction

The line intensity in class 1 samples is a function of both the concentration of the element being determined and the over-all sample composition [see equation (1)]. Therefore, for fluorescent X-ray spectrographic analysis to be quantitative, either the standards and unknowns must be similar or some correction must be applied for matrix effects (variation in sample composition).

High-purity metals, oxides, and simple solutions consist of $99 + \%$ of the major element; therefore, matrix effects will not be a problem. Samples of variable compositions such as ores or minerals will require some type of external or internal standard to correct for differences in composition between sample and standard. Fortunately, the various techniques developed for major and minor constituents [10,44] are equally applicable in the trace-element range.

Class 2 samples generally do not require matrix corrections because standards are prepared and treated in the same manner as the unknown. In many instances a standardization step may be included in the technique; for example, a second element, added to coprecipitate with the element being determined, will serve both as a carrier and an internal standard.

## Theoretical Limit of Detection

Other investigators [25,45,59] have shown fluorescent X-ray spectrographic intensities to be a random process similar to radioactive decay. Therefore, it is possible to predict the theoretical limits of detection for a given confidence level, assuming instrumental variations are negligible. These theoretical limits are obviously very important as they are the best that can be achieved under present conditions. In order to lower these theoretical limits of detectability, advances in instrumentation or techniques will be required.

In the trace-element range, $N_T - N_B$ is proportional to the concentration of element $A$ where $N_T$ is the total number of counts at peak position with element $A$ present for time $t$ and $N_B$ is the total number of counts at peak position with element $A$ absent for the same time $t$.

The standard deviation (counting error) for $N_T - N_B$ is equal to $\sqrt{\overline{N}_T + \overline{N}_B}$ where $\overline{N}_T$ and $\overline{N}_B$ are the arithmetic means of the $N_T$ and $N_B$ terms, both having a Gaussian distribution. In the limiting case the counting error approximates $\sqrt{2}\ \sqrt{\overline{N}_B}$ as

$$\lim_{t \to 0} \overline{N}_T = \overline{N}_B$$

In this paper the minimum detectable amount is defined as that concentration or amount that results in a line intensity above background equal to three times the square root of the background for counting times that are not to exceed 10 min. This gives a confidence level of 95% as $3\sqrt{N_B}$ approximates $2\sqrt{\overline{N}_T + \overline{N}_B}$ at the trace level. At the minimum detectable limit (based on a 95% confidence level or $2\sigma$), $N_T$ will be greater than $N_B$ 39 out of 40 times, and 1 time out of 40, $N_T$ will be less than $N_B$. A more detailed discussion of the minimum detectable amount is given by Zemany [58].

Longer counting times would result in a lower theoretical limit of detectability; however, this limit is a function of the square root of the total number of counts, *e.g.*, increasing the counting time by a factor of 10 will lower the limit of detectability by 3.1. As there are practical limitations which must also be considered, a realistic counting time should be used for each analysis.

At a concentration level of ten times the minimum detectable amount, determinations can be reported to $\pm 10\%$ of the amount present at a 95% confidence level. When $N_T$ and $N_B$ differ by factors of 2 or more, the optimum division of the total time for each measurement should also be considered [48].

## CLASS 1 SAMPLES

As stated earlier, the principal objective of this paper is to summarize applications of fluorescent X-ray spectrographic analysis to

trace analysis. The reader should be aware that the limits of detectability reported by various investigators are not necessarily the ultimate that could have been reached. The lower limits stated were
often those required to meet the demands of the assigned problem;
lower values were, therefore, unnecessary. Also, various definitions
of limits of detectability were used, not all of which were based on
probability theory.

## Metals

National Bureau of Standards ingot iron and low-alloy steel
standards, series 1161–1179, were used to determine the limits of
detectability in high-iron samples. These samples, which are prepared for use as X-ray spectrographic standards, are certified for 15
elements covering a wavelength range of approximately 0.4 to 7 A.
For samples of this type, the summation of mass absorption coefficients [see equation (1)] is independent of changes in the concentration of the impurity element over a range of 0 to 1%; thus, a linear
relationship between $I_A$ and $W_A$ results. The samples were run as
received in a Philips Electronics three-position inverted X-ray spectrograph with an FA-60 tungsten-target tube as the source of excitation. Line intensities, background, and calculated limits of detectability are given in Table III. Intensity values were read off the
analytical curve at the 1000 ppm level; background values were
obtained by extrapolation to zero concentration. Large metal samples
of the 1161 series type can be prepared in a uniform manner; thus,
the practical limits of detection should closely approximate the
theoretical values.

Addink [1], using British Chemical Standard Steel samples, reported similar results with, generally, a slightly lower sensitivity
for most of the elements, but did not give details of the instrumentation used. Bogler [5] gave the sensitivity for various elements in
steels to be 10 ppm for the heavier elements and 100 ppm for the
lighter elements (low atomic number), and also confirmed earlier
findings by Campbell [11] in regard to analytical applications of
long-wavelength $L$ series spectral lines.

Beryllium metal samples were used to estimate limits of detectability in metals having low X-ray absorption characteristics. Calculated intensities (Table I) indicated that sensitivity should be
substantially increased as compared with a metal sample composed of
elements of higher atomic number. Metal powder samples consisted

## TABLE III
### Theoretical Limits of Detectability in Iron

| Element and spectral lines | Instrumentation,* 56 kv, 40 ma | Intensity, counts/sec[†] | | Limit of detection, ppm[‡] |
|---|---|---|---|---|
| | | Line | Background | |
| Si $K_\alpha$ | A | 1.5 | 4.2 | 170 |
| Ti $K_\alpha$ | B | 817 | 80 | 1.0 |
| V $K_\alpha$ | B | 1095 | 286 | 1.9 |
| Cr $K_\alpha$ | B | 1400 | 2130 | 4.0 |
| Mn $K_\alpha$ | B | 1080 | 150 | 1.4 |
| Ni $K_\alpha$ | C | 150 | 43 | 5.4 |
| Cu $K_\alpha$ | C | 142 | 97 | 8.5 |
| As $K_\alpha$ | C | 130 | 52 | 6.8 |
| Zr $K_\alpha$ | C | 298 | 126 | 4.6 |
| Mo $K_\alpha$ | C | 328 | 142 | 4.5 |
| Sn $L_\alpha$ | B | 223 | 50 | 3.9 |

*A = EDDT crystal, helium, 0.02- by 4-in. collimator, flow-proportional counter with PHD.
 B = LiF crystal, helium, 0.02- by 4-in. collimator, flow-proportional counter with PHD.
 C = LiF crystal, air, 0.005- by 4-in. collimator, scintillation counter with PHD.
†Concentration = 1000 ppm.
‡Limit of detection = concentration that results in a line intensity equal to three times the square root of background for 10-min counting time.

of 99+% beryllium.* These samples, NBL 85 to 88, have tentative values of ten elements in the X-ray spectrographic range of atomic number 12 and above; all values are based on chemical analysis.

Three-gram samples of 100-mesh beryllium were used because only a limited amount was made available. Instrumentation was similar to that used to obtain the data for Table III. Larger samples should give improved sensitivity as 3-g samples do not correspond to an infinitely thick sample, except for $Mg K_\alpha$, $Al K_\alpha$, and $SiK_\alpha$ radiation.

For elements of higher concentrations, 100 to 2500 ppm such as iron, nickel, and manganese, the intensity-to-concentration relationship was linear with little scatter of data. In contrast, the data for molybdenum and cobalt were widely scattered, possibly because of larger relative errors in the chemical analysis in the 1- to 30-ppm range.

Intensities for the 100-ppm level for various elements (Table IV) were estimated from the analytical curves; background values were obtained by extrapolation to zero concentration and also at the 1°

*Obtained from Dr. C. J. Roddin of the AEC laboratories, New Brunswick, N. J.

## TABLE IV
### Theoretical Limits of Detectability in Beryllium

| Element and spectral lines | Instrumentation* | Intensity, counts/sec† | | | Limit of detection, ppm‡ |
|---|---|---|---|---|---|
| | | Line | Background | Background 1°2θ off peak | |
| Cr $K_\alpha$ | A, 20kv, 15ma | 220 | 360 | 60 | 1.0 |
| Ni $K_\alpha$ | A, 20kv, 15ma | 300 | 270 | 170 | 0.5 |
| Ni $K_\alpha$ | B, 55kv, 15ma | 520 | 420 | – | 0.5 |
| Co $K_\alpha$ | B, 55kv, 15ma | 400 | 40 | – | 0.2 |
| Mn $K_\alpha$ | A, 20kv, 15ma | 460 | 300 | 100 | 0.5 |
| Mn $K_\alpha$ | B, 55kv, 15ma | 80 | 40 | 14 | 1.0 |
| Fe $K_\alpha$ | B, 55kv, 15ma | 165 | 350 | – | 1.4 |
| Fe $K_\alpha$ | A, 15kv, 6ma | 60 | 100 | 25 | 2.0 |

*A = LiF crystal, helium, 0.02- by 4-in. collimator, flow-proportional counter with PHD.
 B = LiF crystal, air, 0.005- by 4-in. collimator, scintillation counter with PHD.
†Concentration = 100 ppm.
‡Limit of detection = concentration that results in a line intensity equal to three times the square root of background for 10-min counting time.

$2\theta$ off-peak position. The high background at the peak position for many elements e.g.,Cr $K_\alpha$, is due to spectral impurities originating from the X-ray tube. Intensity values for aluminum and silicon, not included in Table IV, indicated a probable detection limit of 100 to 200 ppm.

Davis and Shalgosky, in a meeting summarized by Garton and Davis [28], reported limits of detectability in beryllium for chromium, manganese, iron, cobalt, nickel,.calcium, and zinc of 30, 17, 10, 6, 5, 30, and 7 ppm, respectively. Their limit of detection, based on scanning, was that concentration which would give a peak of 2 divisions on a chart of 100 divisions. Their values are, therefore, significantly higher than those given in Table IV. Even under present limitations, fluorescent X-ray spectrography appears to be very valuable for trace analyses in beryllium metal.

Lambert [42] stated that the minimum measurable amount of uranium or plutonium in aluminum was 500 ppm. For plutonium the sample was sandwiched between two $^1/_4$-mil Mylar sheets to prevent contamination of the sample chamber. These limits seem high since aluminum is not a strong X-ray absorber for radiation in the 1-A range.

## Oxides

In many instances it may be necessary to convert metal samples of irregular shapes or sizes to oxides before analysis. The intensity of a metal $K_\alpha$ line should be approximately the same whether measured on a pure metal or its oxide, assuming an infinite thickness of sample. The intensity per metal atom and the number of metal atoms excited would be about the same in each case. The greater depth of penetration into the oxide is counteracted by its decreased density of metal atoms. Sample preparation is somewhat more critical as the oxide surfaces cannot be polished as evenly or as uniformly as metal surfaces.

Heinrich and McKinley [31] reported limits of detectability in niobium pentoxide of 100 ppm for tantalum and zirconium, 50 ppm for iron, and 30 ppm for cobalt, chromium, manganese, nickel titanium, and vanadium. These samples were diluted 1 : 9 with borax and then fused to give a homogeneous sample.

Barstad and Refsdal [3], using line intensities 10% above background as their criteria of detectability, gave limits of 41, 11, 21, 28, 36, 61 ppm for chromium, nickel, arsenic, strontium, erbium, and thallium in calcium carbonate, respectively. More realistic limits, based on counting statistics, would result in significantly lower values.

Hess [33] established limits of detectability of 100 to 500 ppm for various rare earths in yttrium oxide, using scanning techniques, and stated that a fourfold increase in sensitivity could be expected if counting techniques were used. Lytle and Heady [47] concluded that fluorescent X-ray spectrography is useful for determining rare-earth impurities in any rare-earth oxide. Individual sensitivities of approximately 100 ppm were attained; sensitivity varied depending on the element sought and the host oxide.

Table V lists detection limits for iron, molybdenum, and tin in tungsten trioxide. These values were determined from 10-g samples prepared by adding known amounts of impurities as oxides to tungsten trioxide. The powders were lightly packed into holders normally used for solutions [16].

## Aqueous and Organic Samples

Line intensities for equivalent amounts of trace constituents are higher for aqueous and organic samples than for metal or oxide

## TABLE V
### Theoretical Limits of Detectability in Tungsten Trioxide

| Element and spectral line | Concentration, wt-% | Intensity, counts/sec* | | Limits of detection, ppm† |
|---|---|---|---|---|
| | | Line | Background | |
| Fe $K_\alpha$ | 0.14 | 116 | 12 | 5 |
| Mo $K_\alpha$ | 0.13 | 142 | 120 | 12 |
| Sn $K_\alpha$ | 0.16 | 120 | 200 | 23 |

*55 kv, 25 ma, LiF crystal, air, 0.005- by 4-in. collimator, scintillation counter with PHD.
† Limit of detection = concentration that results in a line intensity equal to three times the square root of background for 10-min counting time.

samples (beryllium excepted). However, scattered radiation is also increased, particularly the ratio of modified to unmodified scattering. Fortunately, this scattered radiation can be reduced significantly by pulse-height discrimination in the wavelength range above 1A. Since the sample surfaces can be readily reproduced, practical limits of detection should approach the theoretical values.

Line intensity, background, and calculated limits of detectability for various elements in water are given in Table VI, using instrumentation similar to that described for Table III. The theoretical values are based on 2-min counting times; longer times would be justified for increased detectability. The improved line-to-background ratio that results from using the longer-wavelength $L$ series line for elements with atomic numbers from 42 to 60, in place of the higher-energy $K_\alpha$ lines, should be noted.

Sladky [52] estimated the lower limits of detectability in nitric acid for various elements between atomic numbers 25 and 50 to be in the range from 20 to 50 ppm. His values are based on the "lowest estimated percent abundance that could be distinguished from 0% by a single measurement at the 95% confidence level."

Barstad and Refsdal [3] give limits of 12 to 44 ppm for strontium, niobium, erbium, uranium, and chromium, and also reported 3 to 6 ppm for nickel, arsenic, and thallium in aqueous solution. A line intensity 5% above background was used as the criterion for limit of detection.

Jones [37,38] found that the sensitivity for uranium and thorium in 1N nitric acid solutions was 5 to 10 ppm. There was some increase in intensity and line-to-background ratio when molybdenum-target tubes were used instead of the usual tungsten target. The intensity of

## TABLE VI
### Theoretical Limits of Detectability in Water

| Element and spectral lines | Instrumentation* | Intensity, counts/sec† | | Limit of detection, ppm‡ |
|---|---|---|---|---|
| | | Line | Background | |
| $S K_\alpha$ | A, 25 ma | 2.8 | 2.1 | 140 |
| $Cl K_\alpha$ | A, 25 ma | 6.8 | 2.4 | 62 |
| $K K_\alpha$ | B, 25 ma | 320 | 14.5 | 3.2 |
| $Ca K_\alpha$ | B, 25 ma | 311 | 6.8 | 2.3 |
| $V K_\alpha$ | B, 25 ma | 772 | 19 | 1.5 |
| $Cr K_\alpha$ | B, 25 ma | 1030 | 370 | 5.1 |
| $Fe K_\alpha$ | C, 25 ma | 2000 | 170 | 1.8 |
| $Cu K_\alpha$ | C, 10 ma | 2100 | 350 | 2.4 |
| $As K_\alpha$ | C, 10 ma | 3140 | 460 | 1.8 |
| $Sr K_\alpha$ | C, 10 ma | 6400 | 1300 | 1.5 |
| $Mo K_\alpha$ | C, 10 ma | 8000 | 3300 | 2.0 |
| $Mo L_{\beta_1}$ | A, 25 ma | 18.5 | 11 | 49 |
| $Cd K_\alpha$ | C, 10 ma | 3200 | 4000 | 5.4 |
| $Cd L_{\beta_1}$ | B, 25 ma | 75 | 8 | 10 |
| $I K_\alpha$ | C, 10 ma | 1200 | 2800 | 12 |
| $I L_\alpha$ | B, 25 ma | 232 | 13 | 4.2 |
| $Ba K_\alpha$ | C, 10 ma | 1070 | 3200 | 14 |
| $Ba L_\alpha$ | B, 25 ma | 275 | 23 | 4.8 |
| $La K_\alpha$ | C, 10 ma | 695 | 2540 | 20 |
| $La L_\alpha$ | B, 25 ma | 275 | 29 | 5.3 |
| $Sm L_\alpha$ | B, 25 ma | 440 | 45 | 4.1 |
| $Sm L_\alpha$ | C, 25 ma | 210 | 62 | 10 |
| $Yb L_\alpha$ | C, 25 ma | 675 | 285 | 6.8 |
| $Au L_\alpha$ | C, 10 ma | 690 | 2400 | 19 |
| $Pb L_\alpha$ | C, 10 ma | 1060 | 460 | 5.5 |
| $Th L_\alpha$ | C, 10 ma | 1220 | 850 | 6.5 |

*A = EDDT crystal, helium, 0.02- by 4-in. collimator, flow-proportional counter with PHD.
  B = LiF crystal, helium, 0.02- by 4-in. collimator, flow-proportional counter with PHD.
  C = LiF crystal, 0.02- by 4-in. collimator, scintillation counter with PHD.
†Concentration = 1000 ppm.
‡Limit of detection = that concentration which results in a line intensity equal to three times the square root of the background for 2-min counting time.

uranium from a solution of 20% tributyl phosphate in Soltrol was approximately 1.5 times greater than in the aqueous systems. This increase was expected because hydrocarbons have a lower total mass absorption coefficient than water. More detailed discussions on analysis of aqueous solutions are given in papers by Campbell, Leon, and Thatcher [16], Lambert [41], Sladky [52], and Flikkema, Larsen and Schablaske [26].

There has been an increasing interest in the determination of trace elements in petroleum or petroleum products. In an earlier paper, Davis and Van Nordstrand [20] determined metallic impurities of barium, calcium, and zinc in lubricating oils. They reported that the zinc content could be measured with an accuracy of $\pm 2\%$ at a 50-ppm level.

Dwiggins and Dunning [24] investigated direct fluorescent X-ray spectrographic analysis for traces of vanadium, iron, and nickel in oils. In an earlier paper [23], they reported an Ni $K_\alpha$ intensity of 241 counts/sec above a background of 102 for 34.3 ppm of nickel and for a 10 min counting time for both line and background; the theoretical limit of detectability was approximately 0.2 ppm.

Hale and King [30] give data that demonstrate their ability to detect 0.07 ppm of nickel in oils at a 95% confidence limit. These workers used scattered radiation that was measured 0.01 A off the peak position for both matrix correction and for background. At 0.1 to 1 ppm, counting times of 30 min are required. Practical limits were the same as the theoretical values.

In solid hydrocarbons, Brown [7] found a deviation of $\pm 2$ ppm at the 10-ppm level for iron in waxes; approximately 15 min were required per sample. Garton and Davis [28] determined lead below 1 ppm in terphenyls that were used as moderators and coolants in a nuclear reactor.

## Minerals

Adler [2] summarized applications of various X-ray spectrographic techniques in geochemical investigations. Some elements of particular interest to the geochemist are uranium, thorium, cerium, yttrium, zirconium, and lead; detection limits of 1 ppm are required for these elements. Accurate knowledge of their distribution provides valuable information on the cooling history of a magma.

Van Wambeke [56] made a detailed evaluation of fluorescent X-ray spectrography for geochemical prospecting and appraisal of niobium-bearing carbonatites. Direct methods with no correction for matrix effects were used; the sensitivity found was 5 to 20 ppm for niobium at an accuracy of $\pm 12\%$. Webber [57] presented a general survey of geochemical prospecting, using X-ray spectrographic techniques.

Herzog [32] determined the ratio of strontium and rubidium in lepidolites for geological age determinations. As the intensity ratio and

## TABLE VII
### Limits of Detectability in Feldspar Matrix*

| Element and | Weight, | Intensity, counts/sec[†] | | Limit of detec- |
|---|---|---|---|---|
| spectral line | % | Line | Background | tion, ppm[‡] |
| K $K_\alpha$ | 0.290 | 400 | 4.0 | 4.0 |
| Ca $K_\alpha$ | 0.257 | 290 | 3.5 | 4.5 |
| Ti $K_\alpha$ | 0.0102 | 25 | 10 | 3.5 |
| Fe $K_\alpha$ | 0.047 | 544 | 124 | 2.6 |

*Bureau of Standards sample No. 99 (feldspar).
[†] 55 kv, 35 ma, LiF crystal, helium, 0.02- by 4-in. collimator, flow-proportional counter with PHD.
[‡]Limit of detection = concentration that results in a line intensity equal to three times the square root of background for 2-min counting time.

concentration ratio for Sr $K_\alpha$ to Rb $K_\alpha$ do not significantly differ, no matrix correction was required. The limits of detectability were stated to be 5 ppm for either element; however, Herzog did not use optimum instrumentation for his measurements, so some improvement is anticipated.

Lewis and Goldberg[43] analyzed marine sediments for titanium, zinc, and barium by conventional X-ray methods and reported that the lower limits of detectability were 100 ppm for titanium and barium and 40 ppm for zinc.

Table VII shows the sensitivity for various elements in a feldspar matrix. These data are of particular interest because elements such as potassium, calcium, and titanium are not considered easily detectable. In another paper, the author [11] pointed out that this sensitivity results in part from the low background in the long-wavelength region.

## Miscellaneous

Brandt and Lazar [6] determined trace metals in plants using dehydrated samples. Sensitivities of 3 ppm for cobalt and zinc and 10 ppm for manganese and molybdenum were given.

Dyroff and Skiba [21] reported sensitivities in the 20-ppm range for nickel and vanadium on alumina-base catalysts. More recent developments in instrumentation makes increased sensitivity possible.

Campbell and Shalgosky [8] compared results in the 30- to 60-ppm range which were obtained on strontium in milk-powder ash by neutron activation, flame photometry, and fluorescent X-ray spectrography.

Campbell, Carl, and White [13] investigated application of fluorescent X-ray spectrography to determine germanium in coal ash extensively. Sensitivities of 100 ppm were found, using a Geiger counter. With the more suitable crystals now available, plus proportional counters used in conjunction with pulse-height discrimination, this limit could be reduced to the 10-ppm range.

## CLASS-2 SAMPLES

Although fluorescent X-ray spectrography is generally limited to concentrations above 1 ppm for class 1 samples, the method is very sensitive for isolated microgram quantities of most elements. Liebhafsky and others [44] stated, "It was estimated that the intensity of $Co\,K_\alpha$ generated under practical conditions in a monolayer area (1 cm$^2$) of cobalt atoms might give 133 counts/sec. Such a sample weighs 0.2 $\mu$g. It would give higher counting rates in a spectrograph especially designed for trace determinations."

### Miscellaneous

A number of successful applications of fluorescent X-ray spectrography were made for analysis of microgram quantities, using preconcentrated samples. Rhodin [50], in his studies of thin oxide films stripped from steel samples, reported sensitivities of 0.037, 0.061, and 0.175 $\mu$g/cm$^2$ for nickel, iron, and chromium.

Campbell and Leon [14] combined selective oxidation with fluorescent X-ray spectrography to determine arsenic and antimony in lead in parts per million. These impurities were found to be concentrated in the surface layers after selective oxidation.

Barstad and Refsdal [3] found limits of sensitivity of $10^{-2}$ to $10^{-3}$ $\mu$g for chromium, iron, nickel, selenium, zirconium, dysprosium, iridium, and thorium, based on line intensities 10% above background. Their background measurements were based on empty support films to obtain theoretical limits. As mentioned earlier, it was found advantageous to reduce X-ray scattering by evacuating the sample chamber, thus significantly lowering the background.

Davis and Hoeck [19], using chemical ashing of residual fuels and charging stocks to preconcentrate vanadium and nickel, were able to determine 1.3 ppm of nickel to ±2.3% of the quantity present and 0.4 ppm of vanadium to $\pm 10\%$. The starting sample ranged from 5 to 20 g. The line intensity for $2\mu g$ of nickel was approximately 100 counts/sec, with a background of 150 counts/sec.

Addink [1] analyzed for zinc in blood in the 3- to 10-ppm range by using 5-ml samples of blood, which were then dry-ashed at 490°C; he did not state the lower limits of detectability obtained by this method.

Cavanagh [18] analyzed for six impurities in high-purity iron (niobium, tantalum, hafnium, zirconium, uranium, and thorium) by electrolytic separation of iron (by mercury cathode techniques) followed by evaporation of the residual solution. The evaporated residue was transferred to a Mylar membrane and placed in a fluorescent X-ray spectrograph. A sensitivity of better than 0.5 ppm for a 10-g sample was reported. More recent developments in fluorescent X-ray spectrography instrumentation should make it readily possible to extend the limit to at least 0.05 ppm for 10-g samples, equivalent to 0.5 $\mu g$ for each element.

Kehl and Russell [39] used fluorescent X-ray spectrography in an investigation of oilfield waters as a source of uranium. The uranium was precipitated from solution and then ashed. The authors reported that as little as $10\mu g$ of uranium could be determined, which was equivalent to 0.01 ppm for starting samples of 1 liter.

Hirt, Doughman, and Gisclard [35] determined the heavy-element content of airborne dust by drawing known volumes of air through a glass-fiber filter disk; this disk was used as the sample support in an X-ray spectrograph. The limits of detectability ranged from 0.5 to $8\mu g$ for elements between atomic numbers 23 and 82. As in other examples cited, recent improvements in instrumentation would increase the sensitivities determined by this technique.

### Ion-Exchange Membranes

A number of investigators have used ion-exchange membranes [49] for separation and collection of trace elements in solutions. These membranes also serve as the sample support in the fluorescent X-ray spectrograph. Grubb and Zemany [29] demonstrated that cobalt in concentrations of $1\mu g$/liter (0.001 ppm) could be analyzed by collecting the element on a cation-type membrane and measuring its X-ray intensity. In 1956, Campbell and Carl [12] discussed various possible applications of ion-exchange membranes in fluorescent X-ray spectrography.

**TABLE VIII**

Reproducibility of Zinc Analysis Using Ion-exchange Membranes*

| Membrane | Side | Zinc, $\mu g$ | Deviation, % |
|----------|------|---------|--------------|
| 1 | a | 199 | −0.5 |
|   | b | 196 | −2.0 |
| 2 | a | 203 | +1.5 |
|   | b | 202 | +1.0 |
| 3 | a | 197 | −1.5 |
|   | b | 197 | −1.5 |
| 4 | a | 200 | 0.0 |
|   | b | 201 | +0.5 |
| 5 | a | 203 | +1.5 |
|   | b | 201 | +0.5 |
| 6 | a | 199 | −0.5 |
|   | b | 204 | +2.0 |

*Amberplex C-1 cation exchange membrane, 200 $\mu g$ cobalt added to 0.01N HCl, 24 hr reaction time

Ion-exchange membranes and ion-exchange filter papers provide a reproducible support for microgram quantities, thus making high-precision trace analysis possible. The reproducibility of zinc analysis using a cation-exchange membrane as both the collector and sample support in the 200-ppm range is demonstrated in Table VIII. Total counts for each intensity measurement were 25,600; therefore, most of the deviation in Table VIII can be attributed to counting statistics. The sample holder used in this study held (Fig. 1) the membrane between two sheets of $^1/_4$-mil Mylar; the location for the membrane was marked on the bottom sheet. The design of this holder is not the best for long-wavelength radiation as there are absorption losses in the Mylar sheets. The Mylar also contributes a significant fraction of the total scattered radiation when thin membranes or papers are used.

The ion-exchange membranes and papers studied by the author ranged in thickness from 2 to 30 $\mu$. This thickness does not meet the requirements for infinite X-ray thickness (for example, $I/I° < 0.001$) except for long-wavelength radiation. The relationship of intensity to amount for any element present in a thin film is given by the following equation:

$$\lambda_2 I_a = (K W_a/R)(1 - e^{-R\rho z})$$

$$R = \lambda_1(\mu/\rho)_s + \lambda_2(\mu/\rho)_s$$

(3)

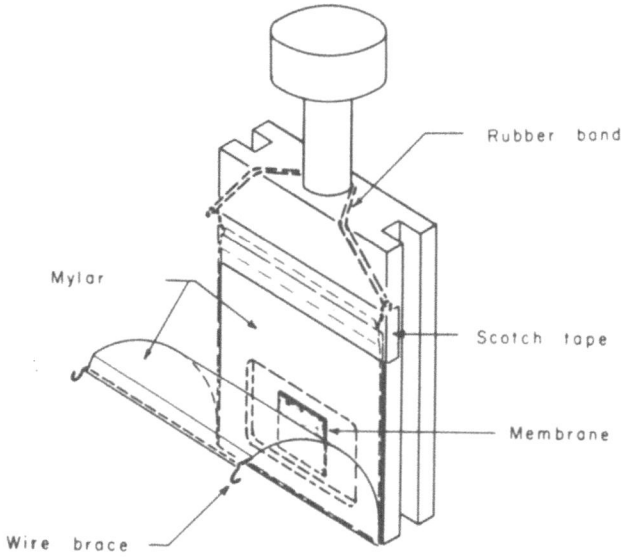

Rubber band

Mylar

Scotch tape

Membrane

Wire brace

Fig. 1. Holder for ion-exchange membrane.

As $x$ increases, $e^{-x}$ approaches zero; therefore, equation (3) becomes equation (1) at infinite thickness.

Figure 2 shows the rapid decrease in transmission when wavelength increases and the thicker Amberplex or Ionics membranes are used; whereas the thinner membranes or papers (Nalfilm, Reeve Angel) can be used efficiently in the long-wavelength region. In general, the thicker membranes have the advantage of greater exchange capacity plus rigidity in the holder, but they limit the sensitivity for radiation that is longer than 2 A.

In dilute solutions the membranes collect the metal ions quantitatively; in more concentrated solutions the equilibrium value is less than 100%. Lytle [46] discusses the problems brought about by limited exchange capacity and incomplete reaction. This problem was also studied in the authors' laboratory. Table IX lists the results obtained for cobalt when the hydrogen ion concentration was varied. In a dilute acid solution, approximately 30% of the cobalt was exchanged after 15 min, whereas only 3 to 4% was exchanged in strong HCl solution. The equilibrium value for cobalt in dilute hydrochloric acid was 84%, whereas only 8 to 9% was exchanged in strong hydrochloric acid solution. More detailed information is available

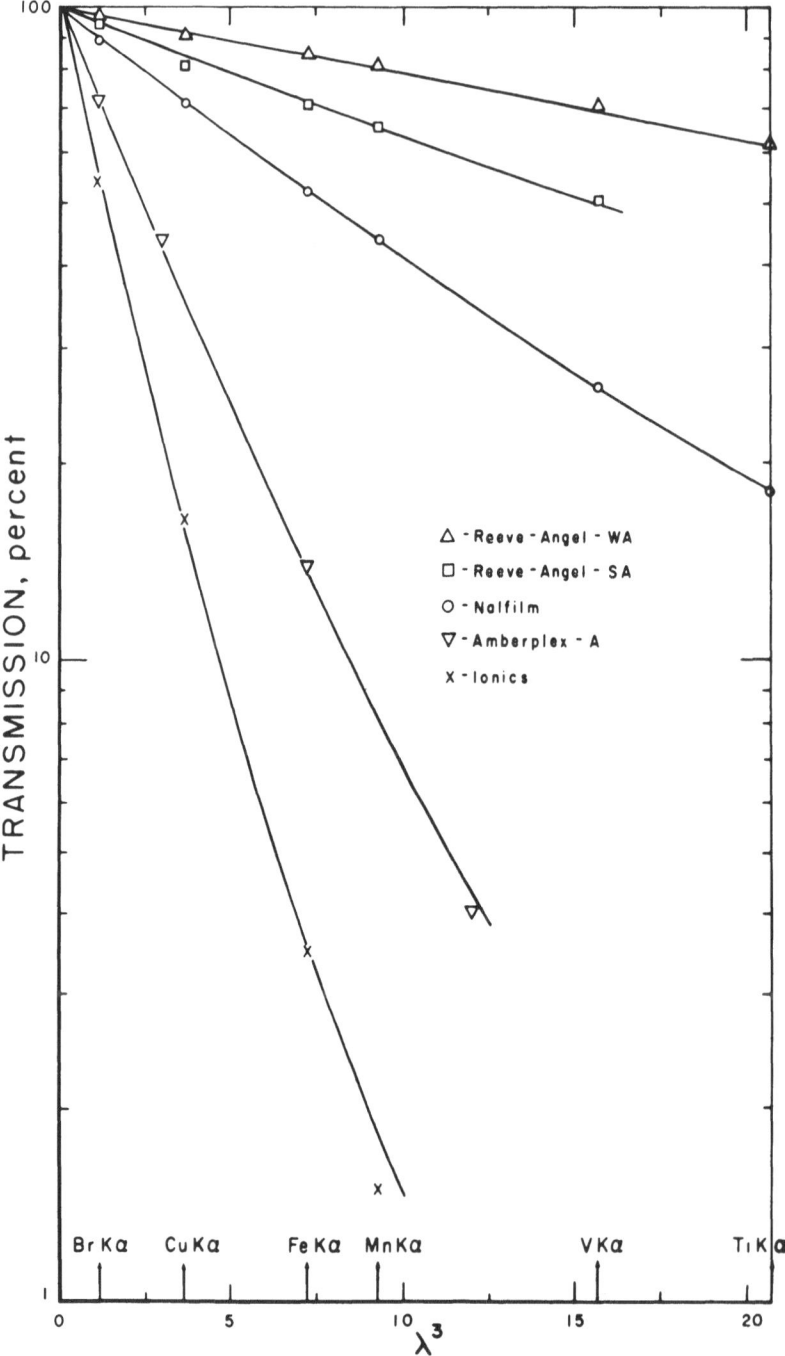

Fig. 2. Transmission coefficients for ion-exchange membranes and papers.

## TABLE IX
### Exchange of Cobalt as a Function of Time
### and Hydrogen Concentration

| Sample* | Reaction time, min† | Cobalt exchanged, % | |
|---|---|---|---|
| | | X-ray‡ | Chemical§ |
| A-1 | 15 | 33.6 | 30.8 |
| A-2 | 30 | 50.6 | 46.8 |
| A-3 | 60 | 68.3 | 64.8 |
| A-4 | 90 | 71.1 | 74.0 |
| A-5 | 120 | 77.2 | 78.0 |
| A-6 | 200 | 80.9 | 82.4 |
| A-7 | 350 | 79.8 | 82.0 |
| A-8 | 500 | 84.9 | 83.6 |
| A-9 | 1500 | 84.0 | 84.0 |
| B-1 | 15 | 3.6 | – |
| B-3 | 60 | 6.8 | – |
| B-6 | 200 | 9.6 | 10.8 |
| B-9 | 1500 | 8.6 | 7.6 |

*A = 25-ml solution with 250 μg of cobalt, 10 ml concentrated HCl per liter.
 B = 25-ml solution with 250 μg of cobalt, 100 ml concentrated HCl per liter.
† Eberback mechanical shaker setting, 55.
‡ Based on chemical value for A-9.
§ Cobalt remaining in solution determined colorimetrically by A. Prokopovitsh, former chemist, Bureau of Mines, College Park.

from numerous textbooks on this subject and from various manufacturers,* some of which are listed below.

The Permutit Company
50 West 44th Street
New York 36, New York

H. Reeve Angel and Co., Inc.
9 Bridewell Place
Clifton, New Jersey

Rohm and Haas Company
Amberlite Division
Philadelphia, Pennsylvania

Ionics Incorporated
152 Sixth Street
Cambridge 42, Massachusetts

National Aluminate Corporation
6218 West 66th Place
Chicago 38, Illinois

Grub and Zemany [29] required extensive periods of time, 24 hr or more, to extract small amounts of cobalt from 1-liter samples.

*Omission of other manufacturers does not imply endorsement of those listed.

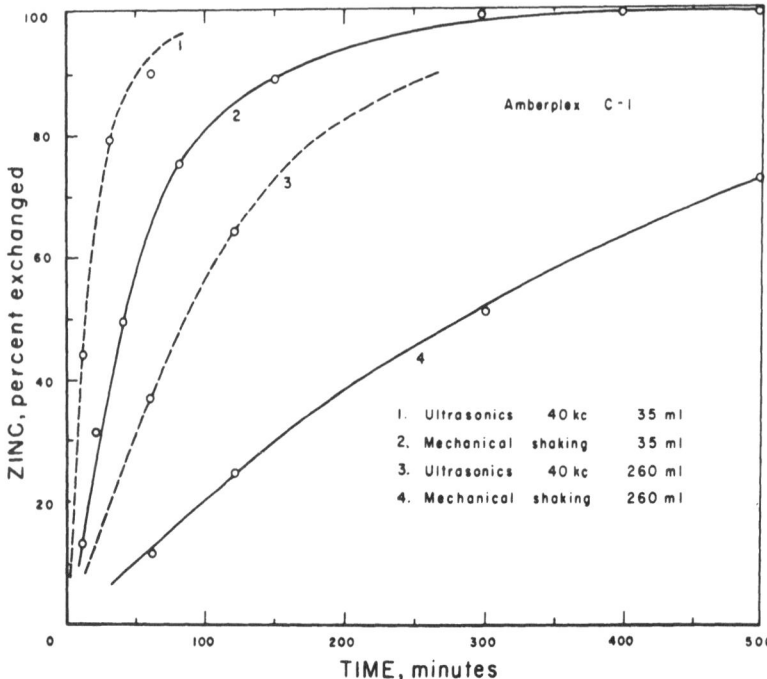

Fig. 3. Variation in exchange rates as a function of volume and type of stirring.

Equilibrium can be obtained more rapidly by using smaller volumes or by increasing the number of collisions of the ions with the membrane per unit time. Data given in Fig. 3 show equilibrium has been approached after 200 min for the 35-ml solution, whereas the 260-ml solution has been 50% reacted. Ultrasonic agitation,* as compared with mechanical stirring, increased the reaction rate by approximately a factor of four. This increased reaction rate may be necessary for applications in process control when time is an important factor. Using powdered or liquid resins, Van Niekerk and De Wet [55] decreased the reaction time to 5 min.

Mutual enhancement and absorption of X-rays by elements in the sample are also observed in ion-exchange membranes. Figure 4 shows the enhancement of the Co $K_\alpha$ line when zinc is added to the solution. These samples were prepared by adding Amberplex membranes, C-1 type, to solutions containing varying amounts of cobalt ions plus 0 to 200μg of zinc. Zinc $K_\alpha$ and $K_\beta$ lines are both

*The authors were assisted in these studies by Charles B. Kenahan, chemist, College Park Metallurgy Research Center, Bureau of Mines.

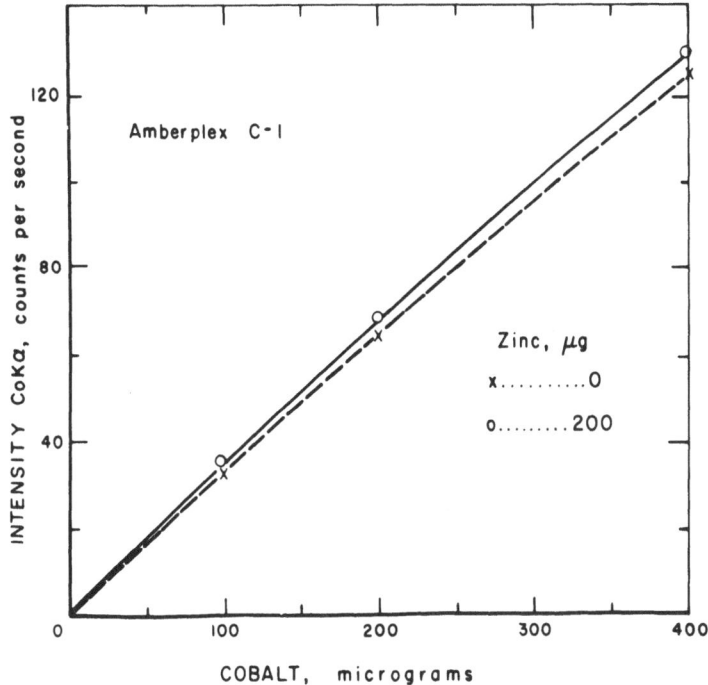

Fig. 4. Enhancement of Co $K_\alpha$ by zinc.

on the short-wavelength side of the Co $K$ edge; thus, they serve as strong sources of excitation. Absorption is demonstrated in Fig. 5; the cobalt ions are strong absorbers of Zn $K_\alpha$ radiation.

## Applications

Van Niekerk and DeWet [55] determined uranium in barren sulfate effluent from ion-exchange columns, using ion-exchange techniques. The major ionic constituents of these solutions were $Fe^{2+}$ and $Fe^{3+}$ (total, 4-6 g/liter), $Mn^{2+}$ (4-6 g/liter), $Al^{3+}$ (4-6 g/liter), and $SO_4^{2-}$ (20 g/liter). By shaking 500 ml of the solution for 5 min with 2 g of an anionic resin, filtering, washing, and drying the resin by suction and then placing it in a suitable holder, 1 ppm of uranium could be determined with an accuracy of ±5%. Limits of detection were extended to 0.1 ppm for solid resins and 0.2 ppm for liquid resins.

Horton and Moak [36] determined microgram quantities of thorium in zircaloy II, using a simplified ion-exchange technique. The zircaloy was dissolved and an ion-exchange membrane which was

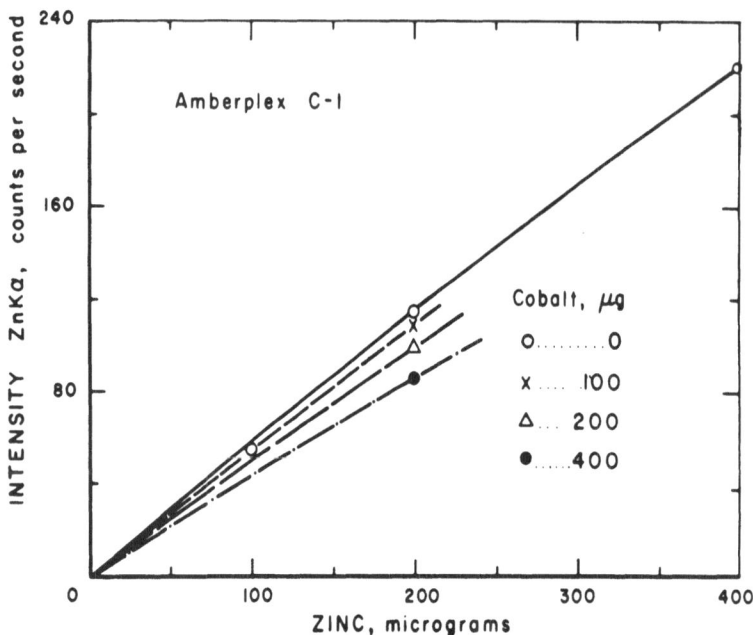

Fig. 5. Absorption of Zn $K_\alpha$ by cobalt.

added to the solution was then washed, dried, and subjected to X-ray analysis. The standard deviation for starting samples of 1 g was ±10% at 5 to 6 ppm.

Campbell, Leon, and Thatcher [16] determined microgram quantities of iron and copper in low-grade copper ores, using cation-exchange membranes. Results obtained by this method and by chemical techniques agreed.

Zemany, Welbon, and Gaines [60] determined the quantity of potassium liberated from the surface of ground-mica samples by ion-exchange-membrane techniques. Their procedure gave a precision of ±15% in the 5- to 150-μg range.

## DISCUSSION AND CONCLUSIONS

A critical evaluation of the literature and the Bureau of Mines experience in this field reveals that fluorescent X-ray spectrography merits serious consideration for analysis of trace elements. For class-1 samples, the limits of detection range from 0.1 to approximately 100 ppm, depending on the element being determined, over-all sample composition, and the complexity of the X-ray spectra.

For class-2 samples, limits of detectability range from 0.01 to 1 $\mu$g, depending on the element being determined and the concentration method used.

The stability of the X-ray tube and detector circuits are extremely important at the lower limits of sensitivity because of the poor signal-to-noise ratio, which can be less than 1 : 200. Multichannel analytical systems do not require critical stabilization of the X-ray-tube output because the monitoring channel compensates for this type of drift. However, electronic drifts in any one of the detector circuits or changes in one of the X-ray optics due to variation in temperature are not corrected in multichannel systems. Highly stabilized single-channel spectrographs should give sensitivities equivalent to those for multichannel units, but twice the counting time is required. This limitation may be important when counting times of $^1/_2$ to 1 hour are required.

Both the reproducibility and accuracy of background measurements are critical; therefore, the technique proposed by Hale and King is recommended [30]. Variations in background intensity due to physical or compositional differences are minimized by making background measurements close to the analytical line. Pulse-height discrimination is desirable since it will limit measured background to X-rays with energies approximating that of the element being determined; multiordered (higher energy) X-rays do not scatter the same as do the lower-energy. radiations; therefore, they should be excluded for the measurements.

When reliable standards are available, calibration of the X-ray spectrograph is comparatively simple because the background contribution can be exactly determined. The authors believe that, in general, such standards are not available; thus, the analyst's individual skill will be very important. The most difficult decision is to determine what fraction of the total count represents the line and the background. This is particularly difficult when some component of the X-ray spectrograph (especially the X-ray tube) contributes spectral lines of the element being determined. The usual practice is to obtain the spectral distribution of the scattered radiation from a high-purity low-atomic number sample, such as a hydrocarbon. This procedure is satisfactory for major or minor constituents, but when trying to attain the ultimate limits of a spectrograph, impurities in the scatterer must also be considered. X-ray tubes free of spectral impurities would be especially useful in this concentration range.

In one of the early papers on fluorescent X-ray spectrography, Freidman, Birks, and Brooks [27] stated that 10 ppm of manganese in aluminum could be determined to approximately ±1 ppm for 10-min counting times; 1 ppm should be detectable for counting times of 1 hr. They go on to state, "It is therefore justified to claim that the type of X-ray spectrograph now available is capable of performing analyses in the range of parts per million under favorable conditions." The results reported in this paper show that their claims were justified.

As usual in a paper of this type, some comments as to possible improvements are included. The authors feel the most promising approach is modification of the excitation process. Although the authors plan to cover this aspect in detail in a subsequent paper, the following is a brief summary of their calculations.

1. Constant-potential power supplies are most effective in the short-wavelength region; the gain is usually less than 2, and there is little improvement in the line-to-background ratio.

2. Higher voltage X-ray tubes permit a gain in intensity; for example, 100 kv permits a 4- to 6-fold gain in intensity for elements around atomic number 50. The calculated gain decreases rapidly as wavelength increases. Improvement in the line-to-background ratio is usually less than a factor of 2 or 3. Obviously, it is possible to excite lines of higher energy, which may have certain specific applications; however, there is the limitation as to suitable diffraction crystals for high-energy radiation.

3. In the long-wavelength region, greater than 3 A, appreciable gains in intensity (20 to 50) are expected when windowless X-ray tubes and selected X-ray-tube targets are used. Calculations by Philips Electronics [53] and by the Bureau agree that this approach should result in a marked increase in intensity.

4. In the 1- to 2-A wavelength range a substantial gain in the line-to-background ratio should be achieved by selective filtration of the primary X-ray beam. Reducing the target-to-sample distance could more than compensate for loss in total primary radiation.

5. Excitation by electrons rather than higher-energy primary X-rays appears to be very promising, particularly in the long-wavelength region. Refractory metals such as tungsten, because of their higher melting point and good electrical and thermal properties, are particularly adaptable to these studies.

# REFERENCES

1. N.W.H. Addink, "The Determination of Trace Elements," *J. Iron and Steel Inst.,* 199–211, 1960.
2. I. Adler, "Application of X-ray Spectroscopy to Unsolved Problems in Geochemistry," *ASTM STP* **269**, 47–54, 1960.
3. G.E.B. Barstad and I.N. Refsdal, "Sensitive Quantitative Recording X-ray Spectrometers," *Rev. Sci. Instr.* **29**, 343–348, 1958.
4. L.S. Birks, *X-ray Spectrochemical Analysis,* Interscience Publishers, Inc., New York, 1959, 137 pp.
5. B. Bolger, "Experiences with an X-ray Fluorescent Vacuum Spectrograph." *5th Internat. Inst. and Measurements Conf.,* Stockholm, Sweden, *Abs.,* 126–127, 1960.
6. C.S. Brandt and V.A. Lazar, "Analysis of Dried Plant Material by X-ray Emission Spectrography," *Agr. and Food Chem.* **6**, 306–309, 1958.
7. J.D. Brown, "X-ray Fluorescence Analysis Applied to Wax Processing and Cellulose Chemistry," *6th Ottawa Symposium on Appl. Spectroscopy,* Ottawa, Can., October 1959.
8. J.T. Campbell and H.L. Shalgosky, "X-ray Spectrographic Determination of Strontium," *Nature* **183**, 1481, 1959.
9. W.J. Campbell, "Application of X-ray Spectroscopy to Trace Element Analysis," *1st Eastern Anal. Symposium,* New York, November 1959, paper 29.
10. W.J. Campbell, "The Determination of Germanium by Fluorescent X-ray Spectrography," Univ. of Maryland, College Park, Md., Ph.D. thesis, 1956, 144 pp.
11. W.J. Campbell, *Fluorescent X-ray Spectrographic Analysis: Studies of Low-Energy K, L, and M Spectral Lines,* Bureau of Mines Rept. of Investigations 5538, 1959, 20 pp.
12. W.J. Campbell and H.F. Carl, "Fluorescent X-ray Spectrographic Analysis of Traces of Germanium," *Pittsburgh Conf. on Analytical Chemistry and Applied Spectroscopy,* Pittsburgh, March 1956, paper 92.
13. W.J. Campbell, H.F. Carl, and C.E. White, "Quantitative Analyses by Fluorescent X-ray Spectrograph—Determination of Germanium in Coal and Coal Ash," *Anal. Chem.* **29**, 1009–1017, 1957.
14. W.J. Campbell and M. Leon, *Fluorescent X-ray Spectrograph for Dynamic Selective Oxidation Rate Studies: Design and Principles,* Bureau of Mines Rept. of Investigations 5739, 1961, 21 pp.
15. W.J. Campbell, M. Leon, and J.W. Thatcher, "Flat Crystal X-ray Optics," *Proc. 6th Conf. on Ind. Applications of X-ray Analysis,* Denver Research Inst., Denver, Colo., August 1957, 193–206.
16. W.J. Campbell, M. Leon, and J.W. Thatcher, *Solution Techniques in Fluorescent X-ray Spectrography,* Bureau of Mines Rept. of Investigations 5497, 1959, 24, pp.
17. W.J. Campbell and J.W. Thatcher, *Determination of Calcium Wolframite Concentrates by Fluorescent X-ray Spectrography,* Bureau of Mines Rept. of Investigations 5416, 1958, 18 pp.
18. M.B. Cavanagh, *The Application of X-ray Fluorescence to Trace Analysis,* Naval Res. Lab. Rept. 4528, 1955, 4 pp.

19. E.N. Davis, and B.C. Hoeck, "X-ray Spectrographic Method for the Determination of Vanadium and Nickel in Residual Fuels and Charging Stocks," *Anal. Chem.* **27**, 1880–1884, 1955.

20. E.N. Davis, and R.A. Van Nordstrand, "Determination of Barium, Calcium, and Zinc in Lubricating Oils—Use of Fluorescent X-ray Spectroscopy," *Anal. Chem.* **26**, 973–977, 1954.

21. G.V. Dyroff and P. Skiba, "Determination of Trace Amounts of Iron, Nickel, and Vanadium on Catalysts by Fluorescent X-ray Spectrography," *Anal. Chem.* **26**, 1774–1778, 1954.

22. C.W. Dwiggins, Jr., "Quantitative Determination of Low-Atomic-Number Elements Using Intensity Ratio of Coherent to Incoherent Scattering of X-rays," *Anal. Chem.* **33**, 67–70, 1961.

23. C.W. Dwiggins, Jr. and H.N. Dunning, "Quantitative Determination of Nickel in Oils by X-ray Spectrography," *Anal. Chem.* **31**, 1040–1042, 1959.

24. C.W. Dwiggins, Jr. and H.N. Dunning, "Quantitative Determination of Traces Vanadium, Iron and Nickel in Oils by X-ray Spectrography," *Anal. Chem.* **32**, 1137–1141, 1960.

25. H.R. Erard and G. L. Underhill, *General Operating Accuracies in Fluorescent X-ray Spectroscopy*, Springfield Armory, Springfield, Mass., SA-TR20-2405, 1959, 53 pp.

26. D.A. Flikkema, R.P. Larsen, and R.V. Schablaske, *The X-ray Spectrometric Determination of Uranium in Solution*, Argonne Nat. Lab. Rept. 5641, Lemont, Ill., 1956, 9 pp.

27. H. Friedman, L.S. Birks, and E.J. Brooks, "Basic Theory and Fundamentals of Fluorescent X-ray Spectrographic Analysis," *ASTM STP* **157**, 3–26, 1954.

**28. F.W.J. Garton and H.M. Davis, "Summarized Proceedings of a Colloquium on** X-ray Fluorescence Analysis," *Brit. J. Appl. Phys.* **10**, 105–116, 1959.

29. W.T. Grubb and P.D. Zemany, "X-ray Emission Spectrography Using Ion Exchange Membranes," *Nature* **176**, 221, 1955.

30. C.C. Hale and W.H. King, Jr., "Direct Nickel Determinations in Petroleum Oils by X-ray at the 0.1 ppm Level," *Anal. Chem.* **33**, 74–77, 1961.

31. K.F.J. Heinrich and T.D. McKinley, "The Determination of Impurities in Elemental Niobium and Its Compounds by X-ray Spectroscopy, *Pittsburgh Conf. on Anal. Chem. and Appl. Spectroscopy*, Pittsburgh, March 1959, paper 61.

32. L. F. Herzog, "Age Determination by X-ray Fluorescence, Rubidium-Strontium Ratio Measurement in Lepidolite," *Science* **132**, 293–294, 1960.

33. T.M. Hess, *Determination of Rare-Earth Oxides in $Y_2O_3$ by Means of the X-ray Spectrograph, Yttrium Anal. Conf., Argonne National Laboratory, Lemont, Ill., Minutes*, November 1956, 76 pp.

34. W.F. Hillebrand, "The Analysis of Silicate and Carbonate Rocks," *Geol. Survey Bull.* **700**, 32, 1919.

35. R.C. Hirt, W.R. Doughman, and J.B. Gisclard, "Application of X-ray Emission Spectrography to Air-Borne Dusts in Industrial Hygiene Studies," *Anal. Chem.* **28**, 1649–1651, 1956.

36. W.S. Horton and W.D. Moak, *Determination of Microgram Quantities of Thorium in Zircaloy II by X-ray Fluorescence Spectroscopy with Ion-Exchange Membranes*, Knolls Atomic Power Lab. Rept. KAPL-M-WSH-4, 1959, 11 pp.

37. R.W. Jones *The Determination of Impurities in Uranium by X-ray Fluorescence Spectrometry*, Atomic Energy of Canada Ltd., Chalk River, Ontario, CRDC–842, 1959, 7 pp.

38. R.W. Jones *Some Applications of X-ray Fluorescence Spectrography to the Determination of Uranium and Thorium*, Atomic Energy of Canada Ltd., Chalk River, Ontario, CRDC–843, 1959, 10 pp.

39. W.L. Kehl and R.G. Russell, "Fluorescent X-ray Spectrographic Determination of Uranium in Water and Brines," *Anal. Chem.* **28**, 1350–1351, 1956.

40. J.W. Kemp, "The Future of X-ray Fluorescence Instrumentation," *ASTM STP* **269**, 55–62, 1960.

41. M.C. Lambert, *Some Practical Aspects of X-ray Spectrography*, Hanford Atomic Products Operation, Richland. Wash., HW–58967, 1959, 66 pp.

42. M.C. Lambert, *X-ray Spectrographic Determination of Uranium and Plutonium in Aluminum Alloys and Other Reactor Fuel Materials*, Reactor Fuel Measurement Techniques Symposium, Michigan State Univ., East Lansing, Mich., TID–7560, June 1958, 208 pp.

43. G.L. Lewis, Jr. and E.D. Goldberg, "X-ray Fluorescence Determination of Barium, Titanium, and Zinc in Sediments," *Anal. Chem.* **28**, 1282–1285, 1956.

44. H.A. Liebhafsky, H.G. Pfeiffer, E.H. Winslow, and P.D. Zemany, *X-ray Absorption and Emission in Analytical Chemistry*, John Wiley and Sons, Inc. New York, 1960, 357 pp.

45. H.A. Liebhafsky, H.G. Pfeiffer, and P.D. Zemany "Precision in X-ray Emission Spectrography," *Anal. Chem.* **27**, 1257–1258, 1955.

46. F.W. Lytle, "Determination of Trace Elements in Plant Material by Fluorescent X-ray Analysis," Univ. of Nevada, Reno, Nev., M.S. thesis, 1958.

47. F.W. Lytle and H.H. Heady, *X-ray Emission Spectrographic Analysis of High-Purity Rare-Earth Oxides*, Bureau of Mines Rept. of Investigations 5526, 1959, 9 pp.

48. M. Mack and N. Spielberg, "Statistical Factors in X-ray Intensity Measurements," *Spectrochimica Acta* **12**, 169–178, 1958.

49. H.G. Pfeiffer and P.D. Zemany "Trace Analysis by X-ray Emission Spectrography *Nature* **174**, 397, 1954.

50. T.N. Rhodin "Chemical Analysis of Thin Films by X-ray Emission Spectrography," *Anal. Chem.* **27**, 1857–1861, 1955.

51. E.B. Sandell, *Colorimetric Determination of Traces of Metal*, Interscience Publishers Inc., New York, 1950, pp. 3–6.

52. R.E. Sladky *Determination of Metallic Impurities in Uranyl Nitrate Solution by X-ray Fluorescence*, Union Carbide Nuclear Co., Y–12 Plant, Oak Ridge, Tenn., Y–1276, December 1959, 25 pp.

53. N. Spielberg, "Intensities of Radiation from X-ray Tubes and the Excitation of Fluorescence X-rays," *Philips Res. Rept.* **14**, 215–236, 1959.

54. N. Spielberg, W. Parrish, and K. Lowitzsch, "Geometry of the Non-Focusing X-ray Fluorescence Spectrograph," *Spectrochimica Acta* **8**, 564–573, 1959.

55. J.N. Van Niekerk and J.F. DeWet, "Trace Analysis by X-ray Fluorescence Using Ion-Exchange Resins," *Nature* **186**, 380–381, 1960.

56. L. Van Wambeke, "Geochemical Prospecting and Appraisal of Niobium-Bearing Carbonatites by X-ray Methods," *Econ. Geol.* **55**, 732–758, 1960.

57. G.R. Webber, "Application of X-ray Spectrometric Analysis to Geochemical Prospecting," *Econ. Geol.* **54**, 816–828, 1959.

58. P.D. Zemany, "The Minimum Amount of an Element Detectable by X-ray Spectrography." *1st Eastern Anal. Symposium,* New York, November 1959, paper 30.

59. P.D. Zemany, H.G. Pfeiffer, and H.A. Liebhafsky, "Precision in X-ray Emission Spectrography—Background Present," *Anal. Chem.* **31**, 1776–1778, 1959.

60. P.D. Zemany, W.W. Welbon, and G.L. Gaines, "Determination of Microgram Quantities of Potassium by X-ray Emission Spectrography of Ion Exchange Membranes," *Anal. Chem* **30**, 299–300, 1958.

# Advances in High-Resolution X-ray Spectrometry

## E. L. Jossem

Department of Physics and Astronomy
The Ohio State University
Columbus, Ohio

Some recent advances in technique and interpretation in high-resolution, precision X-ray spectrometry are reviewed. Topics discussed include improvements in X-ray generation and detection systems, the "thickness effect" in X-ray absorption measurements, temporal changes in the spectrometric properties of analyzing crystals, and computational techniques for correcting spectra for instrumental resolving power and for the effects of inner electron states. Applications of these advances to the investigation of the electron energy band structure of solids, to studies of crystal perfection, and to the establishment of precision X-ray wavelength scales are also discussed.

This review is concerned with recent advances in technique and interpretation in the field of precision, high-resolution X-ray spectrometry. It is, of course, not practicable to consider here in a comprehensive way the various developments in this broad and complex field. One can at best hope to present a bird's-eye view of the subject and highlight a few of its special aspects. References to pertinent literature are, however, included to assist further inquiry. To state the area of discussion more precisely, the review is concerned with X-ray spectrometry in the crystal region ($1 \text{ xu} \leq \lambda \leq 20 \text{ A}$) using instruments which have a numerical resolving power ($R \equiv \lambda/\Delta\lambda$) in the range of $10^4$ to $10^5$, and which are capable of measuring a wide range of X-ray intensities with a precision of $0.1\%$ or better. We shall consider in turn (I) some of the areas in which high-resolution spectrometry is of interest, (II) the general nature of the problems encountered, and finally, (III) a few specific examples of recent work in the field.

I. Precision high-resolution spectrometry is difficult and frequently involves much tedious labor; the incentive for doing it comes from the importance of its applications in the investigation of problems in a variety of fields. For example, in solid-state physics X-ray spectrometry offers unique advantages in the investigation of the electronic energy band structure of solids [1]. Information about the density and wave function symmetry of the filled outer levels in a solid may be obtained from studies of the characteristic X-ray emission lines which arise from the transition of an electron from a filled outer level to a vacancy in an inner level of an atom of the solid. Studies of the structure in the characteristic X-ray absorption spectrum near an absorption edge, arising from electronic transitions from inner levels to the normally unfilled outer levels in the solid, provide complementary information about the density and wave function symmetry of the unfilled outer levels. Physical or chemical changes in the solid perturb the electron energy band structure, and these perturbations are mirrored in the X-ray spectra by changes in wavelength and shape of the emission lines and by changes in the structure of the absorption spectrum. The information about the electronic structure of the valence and conduction bands, both in normal and in perturbed solids, which can be obtained from such studies of the characteristic X-ray spectra is of fundamental importance in the process of understanding the various physical and chemical properties of the solid.

In addition to the studies of characteristic spectra just mentioned, studies of the structure which appears at the short-wavelength limit of the continuous X-ray spectrum [2] and studies of the excitation potentials of emission lines [3] also provide information about the electronic energy band scheme of the material which serves as the X-ray target.

The precise determination of X-ray wavelengths and of the lattice parameters of crystals represent twin aspects of high-resolution spectroscopy, reciprocally related through the Bragg law, which command interest in crystallography, solid-state physics, nuclear physics, and in the determination of precise values of the atomic constants. Crystallographers and solid-state physicists share an interest in precision determinations of the lattice parameters of a crystal and in the study of the effects of imperfections (*e.g.*, dislocations, chemical impurities) on the lattice parameters, physical diffraction patterns, and other surface and volume properties of the crystal.

The nuclear physicist also has need of precise values of X-ray wavelengths in his determinations of the energy levels of the nucleus [4]. The radiation emitted when an excited nucleus makes a transition from one state to another often interacts with the rest of the atom in such a way as to eject an inner-shell electron—a "conversion electron." A measurement of the kinetic energy of such a conversion electron combined with a knowledge of its binding energy in the atom gives the energy of the radiation which was emitted in the nuclear transition, and thus the difference in energy of the two nuclear states involved. Since the binding energy of an inner-shell electron in an atom is determined from the wavelengths of the X-ray absorption edges, the precision with which these X-ray wavelengths are known influences the precision with which the energy levels of the nucleus can be determined. Precise measurements of the wavelength and shape of X-ray lines are also of interest in nuclear physics for the information they contribute to the problem of the finite size of the nucleus and to the study of vacuum polarization and other quantum electrodynamic effects [5].

As a final example of the importance of high-precision, high-resolution X-ray spectroscopy, we may mention the role it has played in the determination of atomic constants, especially in the determination of $h/e$, and in the determination of the conversion constant $\Lambda$ between the grating and the crystal scales of X-ray wavelengths [6,7].

**II.** The general nature of the instrumental problems encountered in high-precision, high-resolution X-ray spectrometry, and in the interpretation of the spectra obtained in such work, ean perhaps be illustrated most easily by a consideration of questions such as the following: What is meant by *the* wavelength of an X-ray line and what is the intensity of the line? What is *the* value of the interplanar spacing for a given set of crystallographic planes in a crystal? What is *the* resolving power of a spectrometer, and how are the distortions of the incident spectrum by the finite resolving power of the spectrometer to be treated? In situations where results of moderate precision suffice, there is little ambiguity about the answers to these questions. The answers become less and less simple, however, as greater and greater precision is required.

When examined in detail, most X-ray emission lines are found to be asymmetrical and many have quite complex shapes. The observed shape of the line and its intensity relative to other lines in the spectrum are affected by the distortions caused by the finite resolving power of the spectrometer and by such other factors as the conditions under which the line is excited, self-absorption in the target, absorption in the X-ray path, the nature of the background radiation which underlies the line, and the stability, linearity, and wavelength sensitivity of the detection system. The precision with which the analyzing-crystal lattice spacing and the Bragg angle are known also enters into the problem of wavelength determination. When angles must be measured to a small fraction of a second of arc, the demands on the alignment, the stability, and the accuracy of the angle-measuring devices become extremely severe. But even assuming that all of the factors mentioned above have been considered and somehow accounted for, the question of what is *the* wavelength of the line remains. Arguments have been presented for and against selecting the intensity peak of the line, the bisectors of horizontal chords, the centroid, *etc.*, but the question is still an open one [6-9]. So also is the question of the theoretical significance of these various measures of wavelength in terms of the fundamental processes of X-ray emission.

As has been mentioned previously, wavelength and lattice spacing are reciprocally related by the Bragg law, and the answer to the question about the lattice parameters of a particular crystal involves many of the factors that have been discussed above in connection with the determination of wavelength. For example, the observed physical diffraction pattern of a crystal depends on the nature of the spectral distribution of the radiation used to observe it. The difficulty of making accurate (as distinct from precise) measurements of lattice parameters is illustrated by the fact that a recent study [10] of lattice parameter measurements made in different laboratories on a selected set of crystal samples has shown that the internal consistency of the measurements made by individual observers may be as much as an order of magnitude better than the agreement among the various observers. Ultimately, the questions of what is *the* wavelength of a spectral line and what is *the* lattice spacing of a crystal need to be faced together, and require also that the concept of resolving power be carefully examined.

In specifying the numerical resolving power $R$ ( $\equiv \lambda/\Delta\lambda$) of a spectrometer one must specify the wavelength interval $\Delta\lambda$. This interval, of course, involves both the geometry of the spectrometer and the physical diffraction patterns of the analyzing crystals. In high-resolution work it is usual to arrange the geometry of the spectrometer so that the diffraction patterns of the crystals are the major factor in the resolving power. It has been usual, also, to take the width at half maximum of the diffraction pattern as determing $\Delta\lambda$. These procedures enable one to calculate a value for $R$ as a figure of merit for the spectrometer. However, as will be seen below, the number so obtained is not necessarily a valid measure of the distortion of the incident spectrum by the spectrometer; in any event, it gives no assistance in the problem of correcting the observed spectrum for such distortion. The difficulty here lies in the fact that a single number is insufficient to characterize the shape of the spectral window of the spectrometer, and that the details of this shape, particularly the extent and relative magnitude of the tail regions, can very strongly influence the nature of the distortion produced.

A more general and more powerful approach to the problem is provided by the convolution or folding equation

$$O(\nu_s) = \int_{-\infty}^{\infty} T(\nu) M(\nu_s-\nu) \, d\nu \tag{1}$$

where $O(\nu_s)$ represents the observed spectrum as a function of the frequency $\nu$, $T(\nu)$ is the "true" incident spectrum, and $M(\nu_s-\nu)$ is the spectral window or measuremental smearing function.* The "true" spectrum, $T(\nu)$, can in principle be obtained by Fourier inversion of equation (1), provided that $O(\nu)$ and $M(\nu_s-\nu)$ are known and that the Fourier transforms of these functions exist.

The convolution equations are, of course, not new. With appropriate changes in terminology one finds them used in a very wide variety of measuremental situations. The point to be especially noted here is that in order to obtain a *unique* solution for $T(\nu)$, one would have to know $O(\nu)$ and $M(\nu_s-\nu)$ with *absolute* accuracy. This condition is never fulfilled in any real situation because of the inevitable presence of noise. In the cases of interest here, the noise results from fluctuations in X-ray intensity caused by fluctuations and drift in the X-ray tube voltage and current and in the intensity distri-

---

* In the case of the two-crystal spectrometer, for example, $M(\nu_s-\nu)$ will include the effects of (a) the geometry of the collimating slits, (b) scattered and fluorescent radiation (especially from the second crystal), (c) higher-order spectra, (d) frequency-dependent absorption in the X-ray path, (e) frequency sensitivity of the detection system, and (f) the diffraction patterns of the analyzing crystals in the antiparallel position.

bution in the focal spot, from fluctuations in the response of the detection system, and ultimately, from the statistical nature of the X-ray emission process and the background radiation. One can never get rid of noise entirely; the best that can be done is to try to minimize its effects by making the signal-to-noise ratio as large as possible.*

Although the difficulties introduced by the presence of noise have long been recognized, relatively little attention has been given to meeting them or to obtaining quantitative estimates of the effects of noise on the accuracy with which a "true" spectrum can be determined. Recently, however, the problem of correction in the presence of noise has been treated in detail specifically for complex X-ray spectra, and an optimized correction procedure has been obtained [11]. The details of the procedure are too complex to permit presentation here, but the method certainly represents the best currently available treatment of the resolving power problem for X-ray spectra.

The discussion of the preceding sections has given some indication of the importance of the spectrometric properties of the analyzing crystals in high-resolution spectrometry. We turn now to a further brief consideration of these properties and of the problems of selecting and of characterizing crystals. Since in high-resolution work the physical diffraction pattern of the crystal usually makes the major contribution to the spectrometer spectral window, one obviously would like to have crystals which have very narrow diffraction patterns with very small tails. When, in addition, one requires that the properties of the crystal be stable with time and that they be stable under exposure to vacuum and X-rays, that the percent reflection be conveniently large, and that the grating space be in a convenient range, then the range of choice of suitable crystals is

---

* The signal-to-noise ratio may be increased either by increasing the intensity of the signal or by reducing the noise, and efforts have been made in both directions. High-power X-ray tubes for fluorescence excitation of spectra (see, e.g. R. J. Liefeld, Bull. Am. Phys. Soc. Series II, 6, 284, 1961) are useful in the reduction of background radiation and in the investigation of materials which are not stable in vacuum or under electron bombardment. Fluctuations in measured X-ray intensity may be reduced by careful attention to the stabilization of X-ray tube voltage and current (0.01% is a reasonable goal) and to the reproducibility of the characteristics of the detection system, especially to the matters of hysteresis and fatigue in the detector. Carefully designed and tested proportional counter and scintillation counter systems are both useful, the balance of the arguments for the one as against the other depending on the individual problem.

very greatly narrowed. Among the natural crystals, calcite, quartz, and beryl have all proved useful, quartz being somewhat more generally useful than calcite, and beryl having its use primarily in the longer-wavelength region out to about 12 A.

The problem of determining the shapes of crystal diffraction patterns is a very difficult one. In the case of the two-crystal spectrometer, one is interested in the diffraction patterns obtaining in the antiparallel (dispersive) positions of the crystals, *e.g.*, the $(1,+1)$ position. Attempts to observe such patterns directly are, however, frustrated by the fact that the finite spectral spread of the incident radiation (*i.e.*, the natural width of the X-ray line used) is comparable with or larger than the crystal pattern width, and thus seriously distorts the patterns. This distortion could be corrected for if one had an accurate knowledge of the *true* shape of the X-ray line, but, as we have seen above, to obtain this true shape one needs to know the details of the spectral window of the spectrometer used to observe it.

Theoretical considerations, such as those of the Darwin-Ewald-Prins theory of diffraction of X-rays by crystals, are of limited help in this problem. Detailed calculations based on this theory have been made for only a few cases, and the assumptions of the theory restrict it to the consideration of the radiation coherently scattered from perfect crystals with identical diffraction patterns. Since incoherently scattered and fluorescence radiation may make a considerable contribution to the tails of the pattern, and since in any practical case one has no *a priori* assurance that the crystals are perfect or that they have identical diffraction patterns, the theory is useful principally in providing a guide as to what one may expect in the ideal case.

One can obtain some indirect information about the crystal patterns from a consideration of the two-crystal spectrometer rocking curves taken in the nondispersive (parallel) position of the crystals, *e.g.*, $(1,-1)$ rocking curves. In this position the dispersion of the spectrometer is zero, or very nearly so, and the wavelength spread of the incident radiation is of little importance. Such $(1,-1)$ curves represent a fold of the single-crystal diffraction patterns of the two crystals. This fold is not the same as that represented by the $(1,+1)$ rocking curves, but it may, nevertheless, be used as a reasonable first approximation to the $(1,+1)$ case, at least as far as the central region of the pattern is concerned.

Recently a study has been made of the spectrometric properties of natural and synthetic quartz [12]. The quantitative aspects of the study are specific for the particular crystals involved, but certain of the qualitative conclusions probably have quite general applicability for crystals used in surface (Bragg) reflection. For instance, the nature of the rocking curves, and inferentially that of the single-crystal diffraction patterns themselves, is critically dependent on the preparation of the reflecting surface. It is possible by very careful selection and surface treatment to obtain crystal pairs with rocking-curve widths which closely approach the theoretical limits. On the other hand, the rocking curves for most crystal pairs, even for those cut from the same piece of raw material, show appreciable asymmetry, indicating that the individual single-crystal patterns are different.

Above the half-maximum ordinate, the rocking curve shapes can be matched well to either a Gaussian or a Lorentzian (classical dispersion) curve. Below the half-maximum ordinate and in the region near the foot of the curves, the rocking curves usually fall somewhere between the Lorentzian and the Gaussian. The tails of the rocking curves eventually cross over and rise above a Lorentzian curve which has been matched at the peak and half maximum, but the position of the crossover points may be asymmetrical for a given pair of crystals and the position of the crossover may vary by a factor of from two to ten for different pairs of crystals. Finally, but by no means least in importance, the width of the rocking curve at the half-maximum ordinate gives no sure indication of the properties of the tails of the curve. Relatively narrow rocking curves may have relatively high and extensive tails and *vice versa*. Curves have been observed which have the same width at half maximum but which have quite different properties with respect to their tails. For precision high-resolution work, therefore, it is necessary to determine in detail the spectrometric properties of the specific crystal pair used. If an appreciable range of wavelength is being investigated, or if the range includes or is in the vicinity of an absorption edge of an element in the crystal, the spectrometric properties must also be determined as a function of wavelength.

III. Turning now to a consideration of some specific work in high-resolution spectrometry, we choose two problems concerned with different aspects of the measurement of X-ray absorption coefficients.

The first of these problems involves the question of where in the X-ray path the absorber should be put when absorption measurements are made with a high-resolution instrument. It has been observed that when such instruments are used, the measured value of the absorption coefficient for certain materials depends on the position of the absorber in the X-ray path. Parratt *et al.* report measurements with a two-crystal spectrometer [13] in which they distinguish three positions: (1) between the source and the first crystal; (2) between the two crystals; and (3) between the second crystal and the detector. The ratio $I/I_0$ of the intensity recorded by the detector with an absorber in the X-ray beam to that recorded with no absorber in the beam was measured for each absorber for each of the three positions. For materials such as aluminum and copper foils, the three $I/I_0$ values were in close agreement, but for graphite blocks and for briquets of polystyrene latex spheres, the $I/I_0$ value in position 2, between the crystals, was lower by a factor of about two to three than the value measured in either of the other positions. The explanation of this phenomenon lies in the very high angular collimation of the X-ray beam provided by the spectrometer, and in the fact that the absorbers for which the effect is large are materials which have a significant amount of small-angle, ultra-small-angle or multiple scattering. The collimation of the transmitted beam, as well as the acceptance range of wavelengths, is essentially determined by the Bragg reflections from the two crystals of the spectrometer. With high resolving power, the angular spread of the collimated beam may be only a few seconds of arc, and radiation undergoing small-angle scattering at angles greater than this will be removed from the beam, thus leading to a lower value for $I/I_0$ in position 2. In position 1 small-angle scattering into the beam largely compensates scattering out of the beam, and in position 3 the angular width of the detector slit is usually great enough to allow the small-angle scattering to enter the detector. The differences in the $I/I_0$ values in the various positions are a measure of the small-scale inhomogeneities of electron density existing in the absorbers, but a detailed interpretation of the relation is difficult in the present stage of development of small-angle-scattering theory. An important point to be recognized here is that the small-angle scattering is a many-atom cooperative phenomenon. If one is making measurements with the intention of obtaining atomic (*i.e.*, per atom) absorption coefficients, then the cooperative scattering intensity should be included in the transmitted beam. Measurements made with the absorber in position 2 effectively exclude this scattering and, thus, will give too high a value for the atomic absorption coefficient.

Our second problem is concerned with the "thickness effect" in the measurement of X-ray absorption spectra near absorption edges [14-16]. The nature of the effect can be seen most readily by examining Fig. 1, which is taken from reference 14. This figure presents the experimental absorption spectrum of chlorine in crystalline KCl near the region of the chlorine $K$ absorption edge. Of the three curves shown, the one with the highest peak at $A$ is for the thinnest absorber, 35,000 A, and the curves with the middle and low values at $A$ are for absorber thicknesses of 140,000 A and 235,000 A, respectively. Over much of the region the three curves are the same within experimental error, but in the region of $A$ and $B$ the differences are striking. This effect too involves the resolving power of the spectrometer, in particular, the effects of the tails of the spectral window in smearing the spectrum.

In practice, curves such as those in Fig. 1 are obtained by

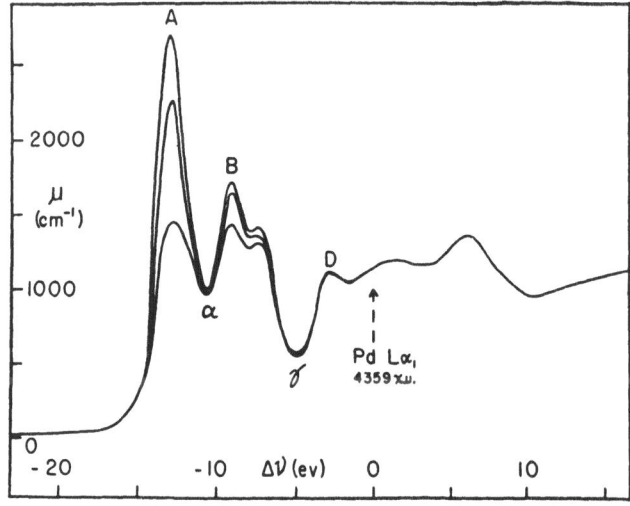

Fig. 1. Experimental absorption spectrum in the region of the $K$ edge of chlorine in crystalline KCl. Three absorber thicknesses were used; in general, the thinner the absorber the greater the absorption coefficient. By the choice of the zero of the ordinate scale, the absorption coefficients refer to the chlorine $K$ electrons only; the measured coefficients are obtained if 640, 410, or 370 is added to the numerical values for the $35 \cdot 10^3$ A, the $140 \cdot 10^3$ A or the $235 \cdot 10^3$ A curve, respectively. (The Pd $L\alpha_1$ line was used as a reference wavelength.) From Parratt, Hempstead, and Jossem, *Phys. Rev.* **105**, 1228 (1957).

determining the transmission spectrum of the absorber, *i.e.*, $I/I_0$ *vs* frequency, and converting the transmission curve to an absorption curve by the usual relation $\ln(I/I_0) = \mu x$. The "Thickness effect" can best be understood in terms of the transmission spectrum, and, by analogy to equation (1), one can write the following expression:

$$O(\nu_s) = k \int_0^\infty I_0(\nu) \, T(\nu) M(\nu_s-\nu) d\nu \tag{2}$$

where here $O(\nu)$ is the experimentally observed transmission spectrum, $k$ is a constant with the dimensions of reciprocal frequency, $I_0(\nu)$ is the incident intensity, $T(\nu)$ is the "true" transmission spectrum, and $M(\nu_s-\nu)$ is the spectral window of the spectrometer. With this in mind we examine Fig. 2, which is also taken from reference 14. In the upper section of this figure are plotted the "true" transmission curves for a thick and for a thin KCl absorber, for the case in which the incident intensity $I_0(\nu)$ is taken to be constant. In the bottom section is plotted the spectral window $M(\nu_s-\nu)$ of the spectrometer as it would appear if the frequency setting dials of the instrument were set at $\nu_s$. Note the great asymmetry of the transmission curves and the fact that the transmission on the low-frequency side of the absorption edge is larger for both absorbers. Remembering equation (2), one can see that although the spectral window may be relatively narrow at half maximum, there will nevertheless, be a considerable contribution to the observed transmitted intensity from frequencies in the range of the window tails, and especially from the tail on the low-frequency side. One can see also that this integrated "leak-through" intensity from the tails will be relatively more important for the case of the thicker absorber and will result in a higher observed transmission ratio and, therefore, in a smaller observed absorption coefficient.

The "thickness effect" is of interest not only as an example of the resolving power problem, but also because, as is discussed in detail in reference 14, it provides another way of obtaining information about the shape and extent of the tails of the spectral window.

Most of the material which has been discussed so far in this section has been concerned with instrumental effects and problems in high-resolution spectrometry. Extremely important advances have also been made in the interpretation of X-ray emission and absorption spectra in terms of the fundamental electronic processes in solids. The principal developments in such interpretation, and in the con-

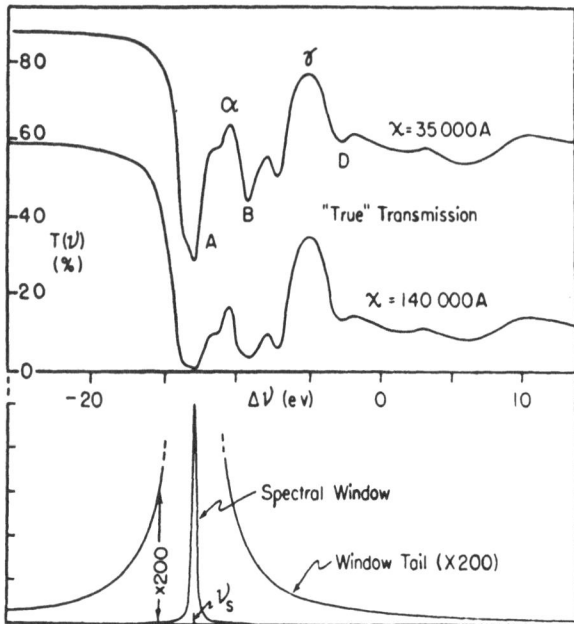

Fig. 2. "True" transmission curves, one for each of two absorber thicknesses, and a typical spectral window. The tails of the window are also shown plotted on an enlarged ordinate scale. The spectral window is pictured with its center $\nu_s$ at the $A$ minimum in the transmission curves. The observed transmission curve $O(\nu_s)$ is obtained by placing $\nu_s$ successively at each of many different positions along the $\nu$ scale and by noting for each position the integral value as expressed mathematically in Eq. (2). From Parratt, Hempstead, and Jossem. *Phys. Rev.* **105**, 1228 (1957).

struction of new types of X-ray energy-level diagrams [1] have sprung from the increasing store of evidence that the one-electron model of processes in solids is inadequate for our present needs. A many-electron model which takes proper account of electron interactions must be used. In this connection, the concept of "excitation states" [1] has proved fruitful in the interpretation of X-ray spectra and has recently received additional experimental support from studies which are concerned with the variations in the shapes of spectral lines observed under special conditions of excitation [17,18]. Unfortunately, the scope and complexity of these ideas prevent their being presented even in brief here, but these last remarks will have served their purpose if they have provoked sufficient interest to send the reader to the original literature.

## REFERENCES

1. L. G. Parratt, *Revs. Modern Phys.* **31**, 616, 1959.
2. R. Sandström, *Arkiv for Fysik* **18**, 305, 1960. Also see P. Johansson, *Arkiv för Fysik* **18**, 329, 1960.
3. A. Nilsson, *Arkiv for Fysik* **6**, 513, 1953.
4. K. Siegbahn [ed], *Beta and Gamma Ray Spectroscopy*, Interscience, New York, 1955.
5. R. L. Shacklett, *Revs. Modern Phys.* **30**, 521, 1958.
6. J. W. M. Dumond, *Proc. Nat. Acad. Sci.* **45**, 1052, 1959.
7. J. A. Bearden and J. S. Thomsen, *A Survey of Atomic Constants*, The Johns Hopkins University, Baltimore, 1955.
8. J. Ladell, W. Parrish, and J. Taylor. *Acta Cryst.* **12**, 253, 561, 567, 1959, and references therein.
9. J. O. Porteus and L. G. Parratt, "Precise Wavelength of a Wide Spectral Line and Precise Lattice Parameter of a Crystal," *Technical Report No. 3*, Cornell University, 1959, AFOSR TN 59-305. ASTIA AD 213 089.
10. W. Parrish, *Acta Cryst.* **12**, 992, 1960 and *Acta Cryst.* **13**, 838, 1960.
11. J. O. Porteus, *J. Appl. Phys.* **33**, 700, 1962, and references therein.
12. E. L. Jossem and P. C. Claspy, *Bull. Am. Phys. Soc. Ser. II.* **6**, 109, 1961.
13. L. G. Parratt, J. O. Porteus, H. W. Schnopper, and T. Watanabe, *Rev. Sci. Instr.* **30**, 344, 1959.
14. L. G. Parratt, C. F. Hempstead, and E. L. Jossem, *Phys. Rev.* **105**, 1228 1957.
15. K. Tsutsumi, M. Obashi, and M. Sawada, *J. Phys. Soc. Japan* **13**, 43, 1958.
16. O. Beckman, B. Axelsson, and P. Bergvall, *Arkiv for Fysik* **15**, 567, 1959.
17. P. Johansson, *Arkiv for Fysik* **18**, 289, 1960.
18. P. O. Schörling, *Arkiv for Fysik* **19**, 47, 1961.

# The Vacuum X-ray Quantometers as Applied to Raw Materials for the Iron and Steel Industries

## B. R. Boyd, H. T. Dryer, and G. Andermann

Applied Research Laboratories
Detroit, Michigan

During 1960 ARL introduced the first commercial multichannel vacuum X-ray spectrometers for the simultaneous analysis of several elements. These instruments have been applied successfully to the analysis of many of the raw materials used in the steel plants and related industries. A report will be presented covering the techniques employed and the results obtained for various iron ores, sinter, slags, and pig iron. Factors relating the speed of analysis to production-control problems will be discussed.

Last year at this symposium I discussed the design concepts of ARL's approach to X-ray instrumentation and the application of vacuum techniques to multichannel X-ray spectrometers. I would like to review briefly some of these instrument concepts and then show how they apply in particular to the analysis of materials for the iron and steel industries.

In the design of X-ray equipment, alternate approaches may be taken to provide the features of the instrument thought to be desirable. Some of the more important of these alternates involve: type of system—monochromator or polychromator; crystal selection—curved or flat, and material; radiation detectors; radiation path—air, helium, or vacuum; and measuring or readout system. Experience in the manufacture of optical emission Quantometers has demonstrated the need for high speed in conjunction with high precision and accuracy for production control. In order to realize these requirements and to insure the use of optimum components for each element, ARL has utilized the parallel or polychromator system of analysis. Some of the advantages of this design system are as follows:

A. *Provides Shortest Analytical Times.* All of these spectrometers or monochromators are used simultaneously; thus, the total analytical time is regulated by the element providing the lowest intensity. The importance of this factor has been emphatically demonstrated as additional light element combinations are required.

B. *Optimum Choice of Components.* Each element and wavelength of interest requires the correct selection of crystal detector, slit system, optical path, *etc.*, to achieve the full potential of X-ray analysis. With the polychromator design, the required system of components can be chosen for any analytical program. Curved crystals of the appropriate material and curvature are used for each element to provide optimum efficiency, resolution, and line-to-background ratio, thereby providing high-quality analytical data.

C. *Provides Simultaneous Ratio of Line to Internal Standard, External Standard, or Scattered Radiation.* The ratio system used depends upon the specific application as previously described by Kemp and Andermann. An additional advantage is gained by this means in that the control channel serves as a monitor control on the X-ray beam intensity, thereby minimizing the need for frequent calibration and increasing the speed of analysis.

D. *Use of Multiple Monochromators.* Although this feature is generally not required, improved speed or precision may be accomplished where required for production requirements. Two or more spectrometers can be set for one element.

Several types of detectors are used to provide the highest efficiency and physical discrimination against high-order resolution for each channel. By a suitable choice of crystal and detector characteristics, physical discrimination is accomplished, permitting a reduction in the complexity of the electronic system. A typical example of physical discrimination by means of the high resolution of the curved crystal spectrometer is shown in Fig. 1. The Si $K_\alpha$ is easily separated from the interfering fourth-order Fe $K_\beta$. The use of the curved crystal provides a focused line, necessitating only a narrow window on the flow detectors and thus reducing gas consumption and the possibility of window rupture.

Three modes of operation are possible with the Vacuum X-ray Quantometers, *i.e.*, air, helium, or vacuum. The air mode would, in

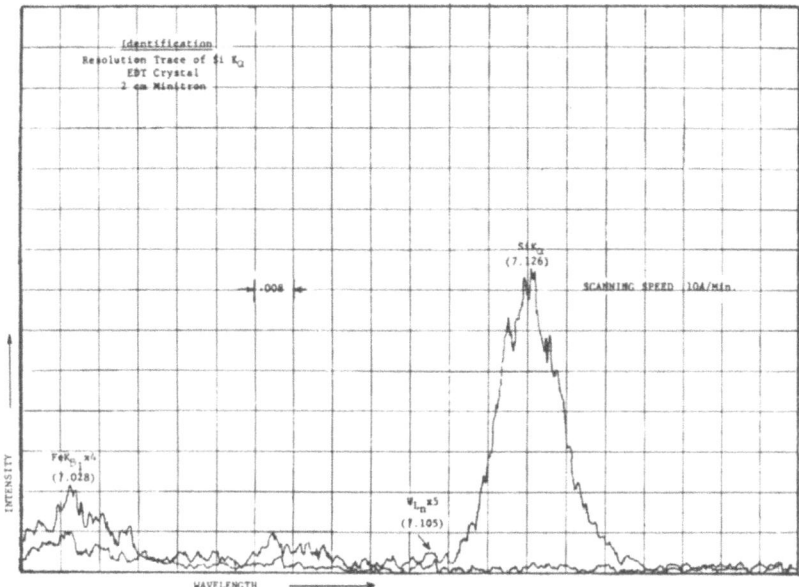

Fig. 1. Physical discrimination by means of the high resolution of the curved-crystal spectrometer.

general, provide suitable data for many applications, *e.g.*, heavy elements in minor or greater concentrations; however, recent studies indicate that vacuum operation provides improved data for elements normally considered heavy elements.

In order to provide high-intensity X-rays for wavelengths longer than about 2.5 A, helium or vacuum paths must be used to eliminate air absorption, which increases with increasing wavelength. The vacuum mode of operation would be preferred to that of helium, both from a cost standpoint and because of the restrictions occasionally imposed on the helium supply. However, because helium would be required for the analysis of volatile liquids for the light elements, all three modes are available and may be used as required. Sample chamber pump-down time for the vacuum mode is of the order of 15–30 sec to achieve a pressure of 0.1 mm Hg. A flush time of approximately 10 sec is used in the helium mode to guarantee high stable intensities.

The detector current integration system, providing a ratio of the element intensity to that of the control channel, is incorporated in the ARL design of X-ray Quantometers. The performance of this type of measuring and readout system has been fully evaluated and demonstrated through its use in optical emission Quantometers and X-ray Quantometers. Reproducibility of the order of 0.1–0.2% is readily available with this measuring system.

Two laboratory types of instruments are currently manufactured by ARL. Both are based on the polychromator design and differ only in the size of the program which can be accommodated and the physical size of the instruments. The VPXQ, shown in Fig. 2, is the larger instrument, capable of handling up to 22 elements simultaneously in an analytical program. The VXQ, a smaller instrument shown in Fig. 3, is capable of handling up to nine elements and, being physically smaller, may be used as a desk unit. In addition to the fixed monochromators, curved-crystal scanning monochromators are available in both instruments to provide added flexibility.

For these two instruments, the VPXQ and the VXQ, the versatility of the scanning monochromators has been combined with the inherent polychromator advantages—speed of analysis and optimum choice of components. The unitized design features of these instruments present the analyst with a choice of expandable instrumenta-

Fig. 2. The VPXQ instrument.

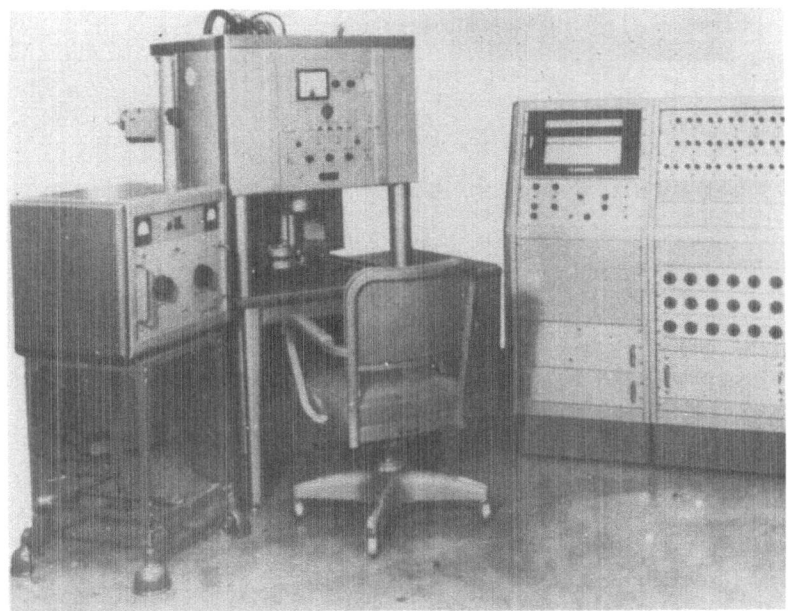

Fig. 3. The VXQ instrument.

tion best suited for his needs in either research or production-control application.

There are many areas of analysis which have succumbed to X-ray fluorescence techniques in the production of iron and steel. The first really valuable application was the analysis of highly alloyed steels such as stainless and tool steels. The next was the handling of some of the ferro- alloys, ferrous manganese and ferrochrome.

I wish to talk specifically about the application of X-ray analysis, both real and potential, to the initial stages of steelmaking, namely, the blast-furnace operations. This phase of steel making has been going through a major overhaul during the past ten years. The tools of automation are beginning to shape up as never before in all areas of product control, and both optical-emission and X-ray fluorescence are playing key roles. Table I shows the various materials requiring rapid analytical control.

Iron ore, the main basic material for the blast furnace, must be available with specific levels of concentration for Fe, Si, and P, since furnace melting control is geared to rather narrow limits. Ore is usu-

## TABLE I
### Blast Furnace Operations

| | Composition, % | | | | | | | |
|---|---|---|---|---|---|---|---|---|
| | Fe | Si | Mn | P | S | Ca | Al | Mg |
| Concentrated ores | 50–68 | 1–15 | 0.1–5 | 0.05–0.5 | | 0.1–2.0 | 0.2–5 | 0.1–1.0 |
| Concentrated taconite | 60–70 | 3–10 | | | | 0.1–1.0 | | 0.1–1.0 |
| Sinter | 50–65 | 2–15 | 0.1–1.0 | 0.01–0.5 | | 2–10 | 0.1–5 | 0.1–5 |
| Blast-furnace slag | 0.1–1.0 | 30–40 | 0.1–2.0 | | | 35–50 | 5–15 | 5–15 |
| Blast-furnace iron | | 0.5–2.0 | 0.2–1.5 | 0.1–0.5 | 0.01–0.2 | | | |

ally concentrated and/or blended at the shipping sites to conform to fairly rigid specifications. The mining of ore and the blending or concentrating depends on the type of ore and the location of the ore body. Within one ore body the mineral combination of the various ore components can vary widely. Therefore, accurate X-ray analysis of ores can be fraught with difficulties. However, recent work in our laboratory shows encouraging results.

Preparation of the sample involves the grinding of all nonmetallic samples in the Bleuler Rotary Mill for 1.5 min with an appropriate material for a binder, and then briquetting with a backing of cellulose powder. The briquets are easy to handle and afford reusable samples for standards. All results shown are based on duplicate runs of 2.5-min instrument analysis time.

Figures 4 through 7 show data on raw ores and one group of fused ores for $SiO_2$ determination. These ores are primarily magnetite and hematite ores from northern Michigan and Minnesota. Figures 8 and 9 cover the concentrated taconite ores. The taconites are low-grade ores which must be concentrated before use. In the concentration process the ore is separated by magnetic means and the resulting product is fairly uniform from a mineralogical standpoint.

In many blast-furnace operations, the ore is sintered prior to introduction to the furnace. The sintering is done to agglomerate the fine ore particles and also to improve the efficiency of the reduction process. The ore to be sintered is sometimes mixed with fluxing compounds, such as dolomite, and is identified as fluxed sinters. Again, as in the case of the low-grade iron ores, the sintered ores are poor specimens as far as mineralogical combinations are concerned and do not provide ideal samples for X-ray analysis in the raw

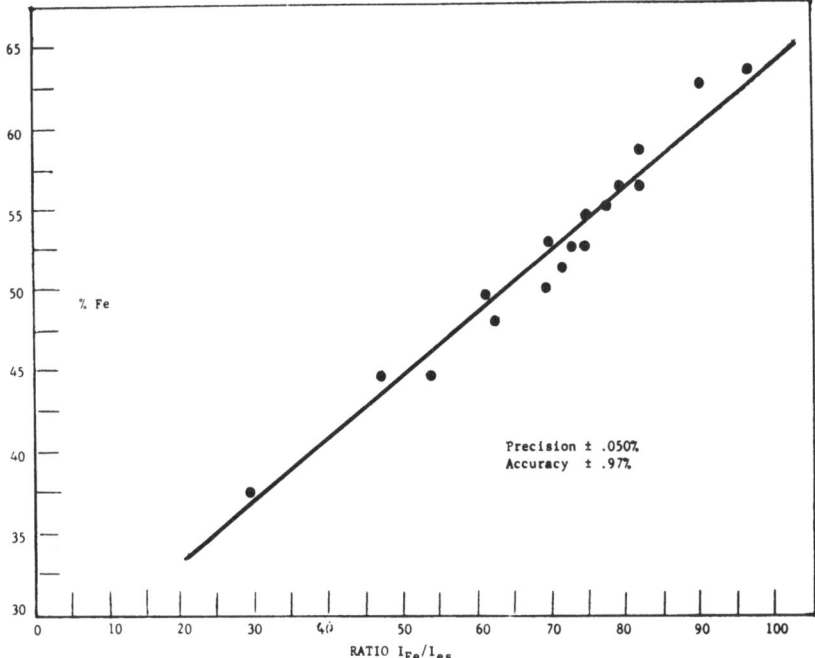

Fig. 4. Fe in iron ore—direct ratio to external standard control $(E/S)$.

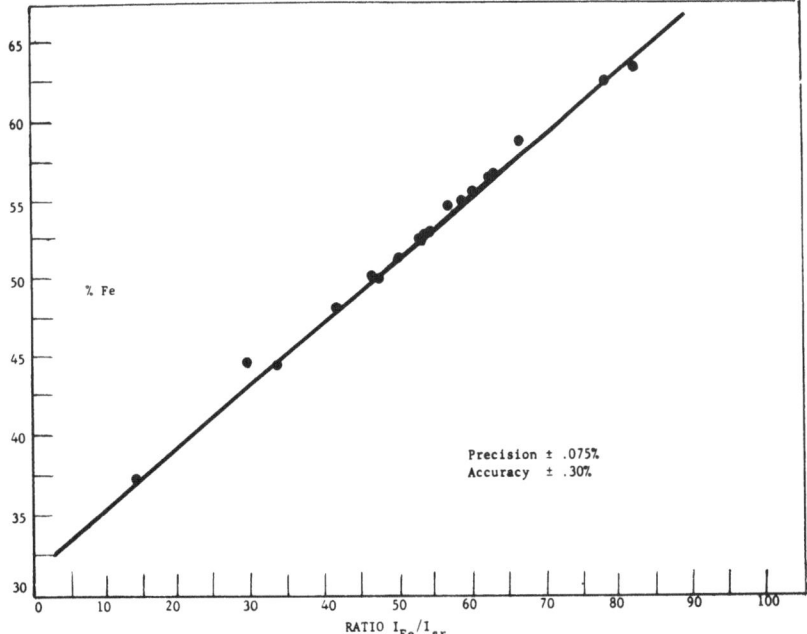

Fig. 5. Fe in iron ore—ratio to scattered radiation from the sample.

B. R. Boyd, H. T. Dryer, and G. Andermann

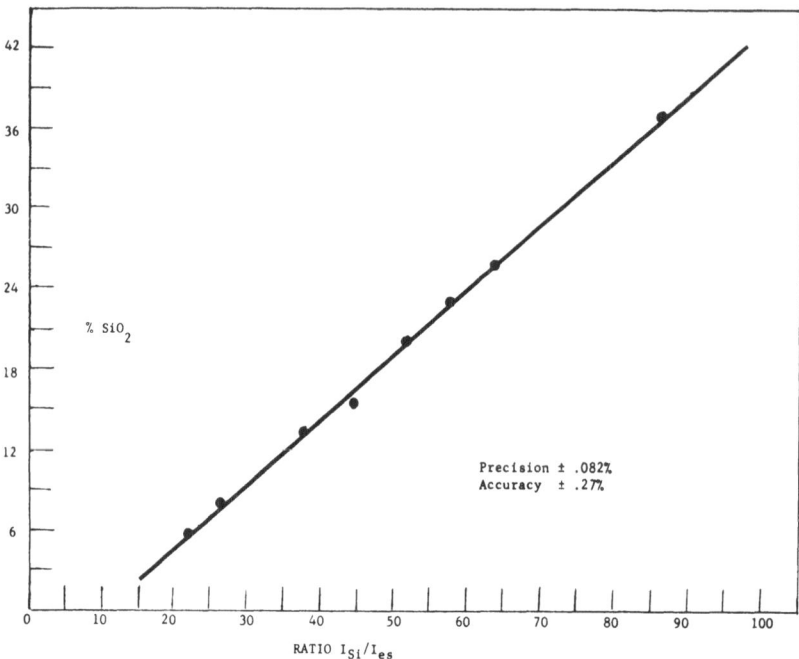

Fig. 6. SiO₂ in fused iron ore samples.

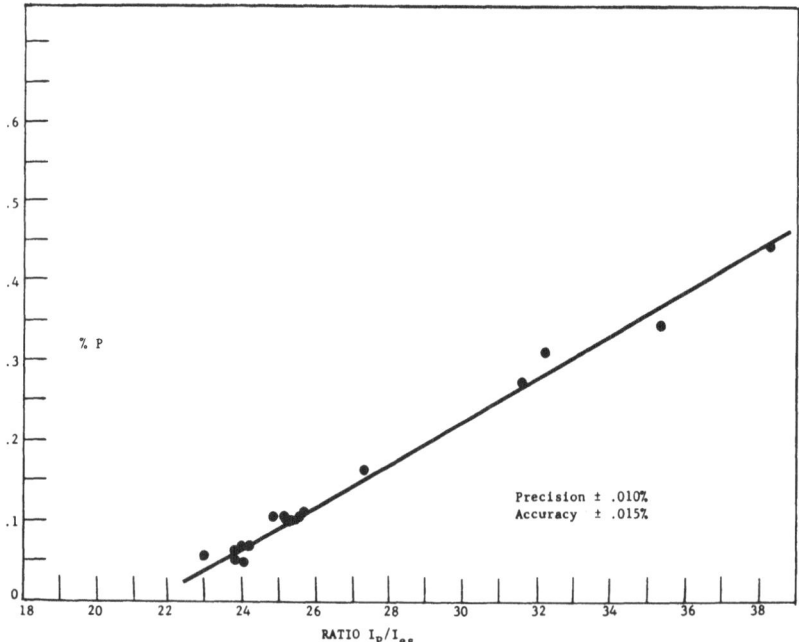

Fig. 7. Phosphorus in raw iron ores.

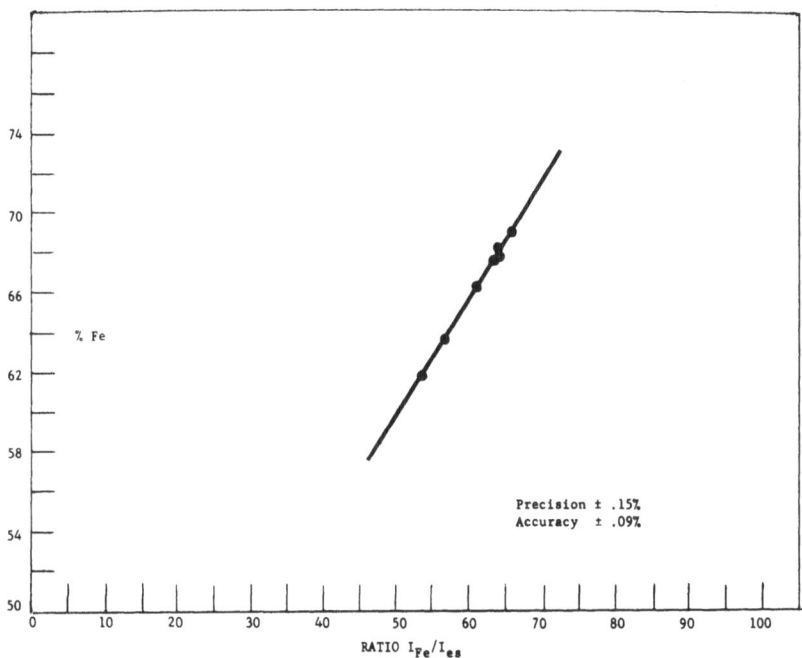

Fig. 8. Fe in concentrated taconite.

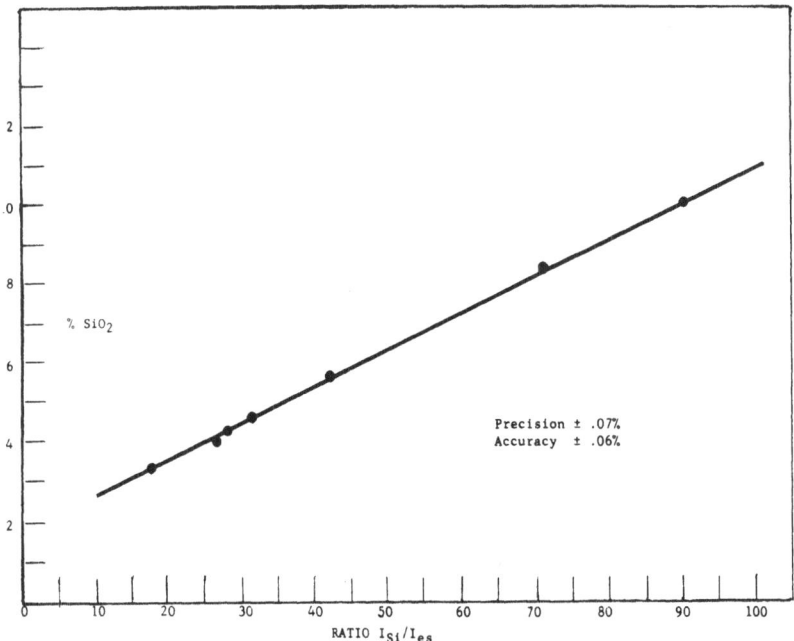

Fig. 9. SiO₂ in concentrated taconite.

**B. R. Boyd, H. T. Dryer, and G. Andermann**

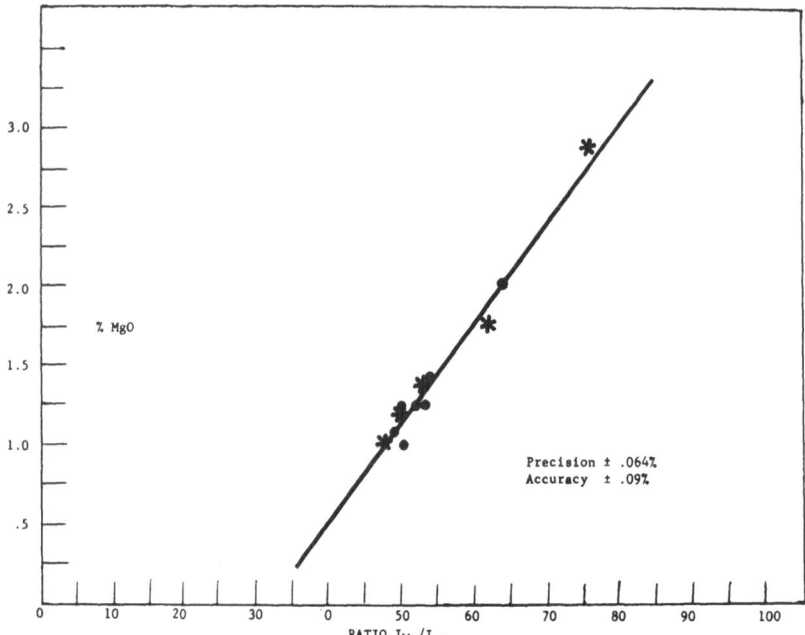

Fig. 10. MgO in sinter.

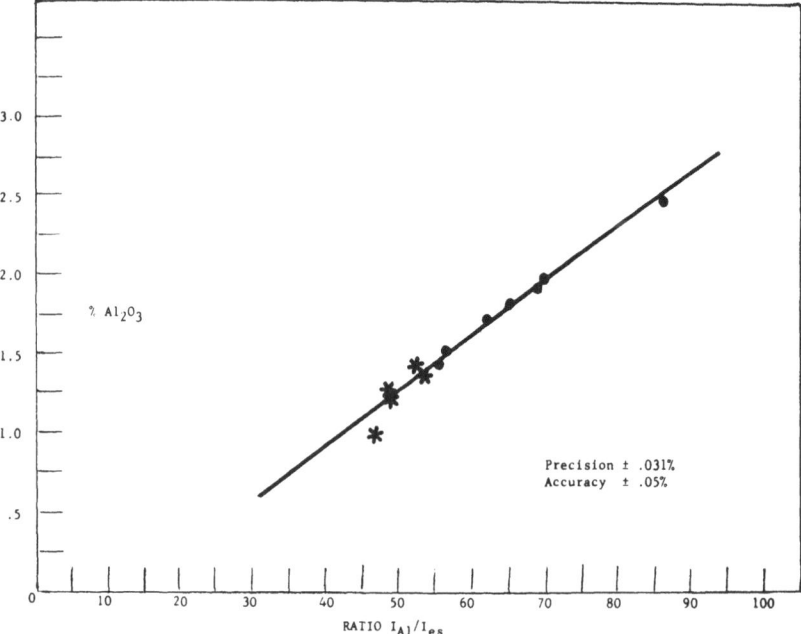

Fig. 11. Al₂O₃ in sinter.

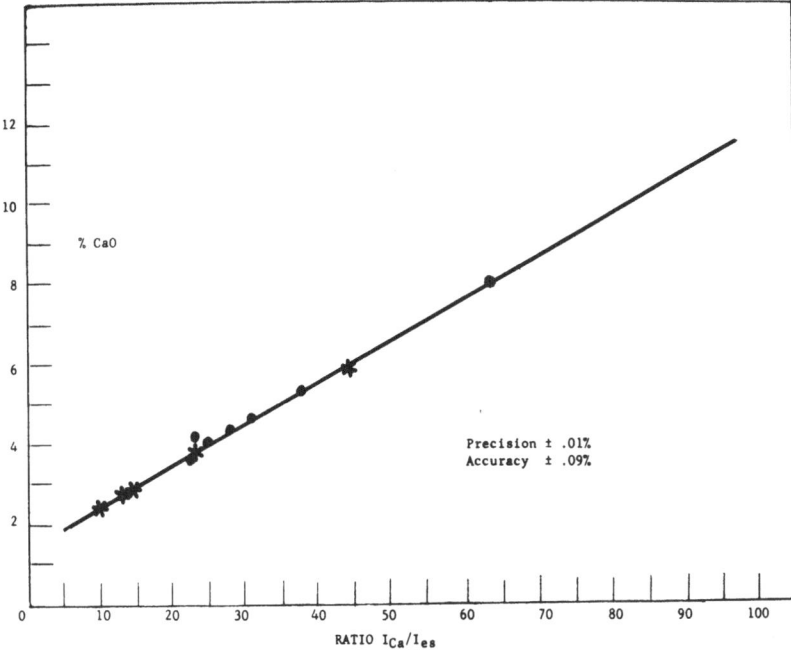

Fig. 12. CaO in sinter.

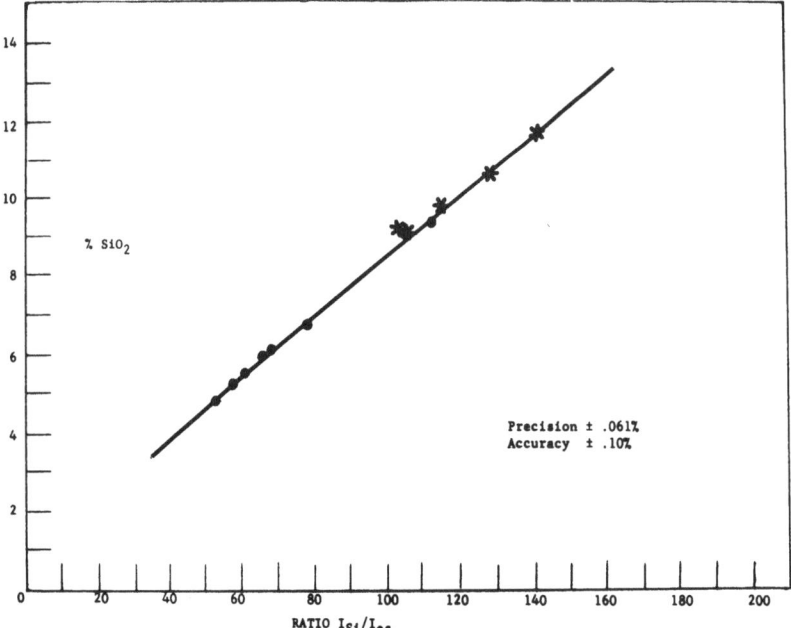

Fig. 13. SiO₂ in sinter.

B. R. Boyd, H. T. Dryer, and G. Andermann

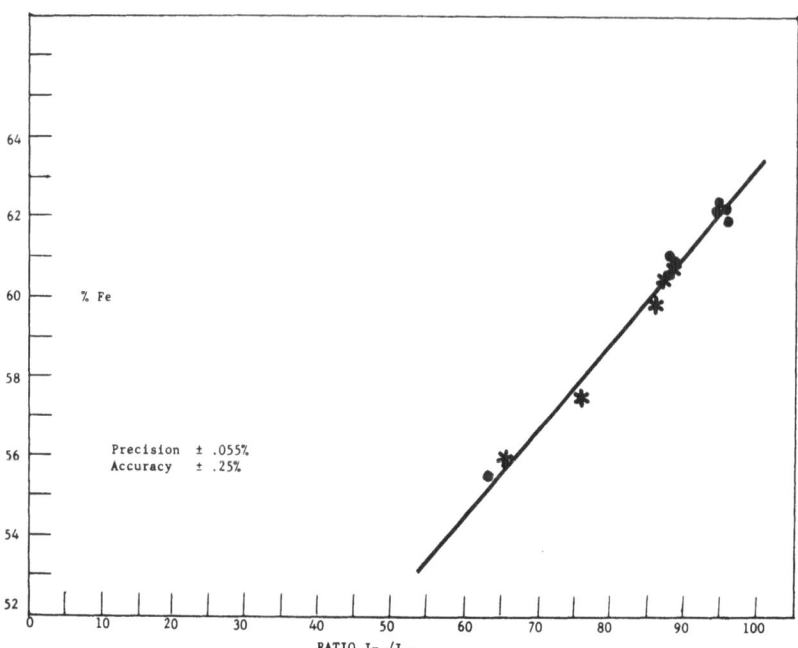

Fig. 14. Fe in sinter.

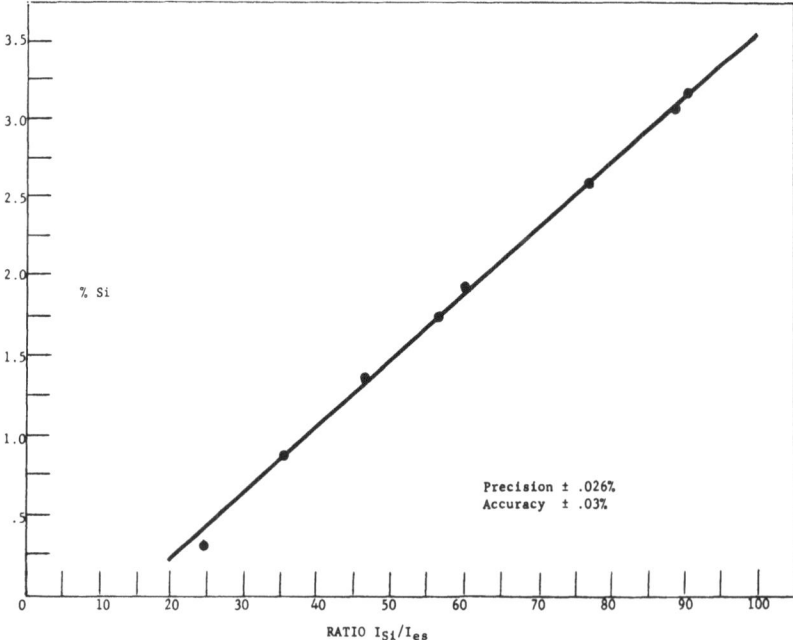

Fig. 15. Si in cast iron.

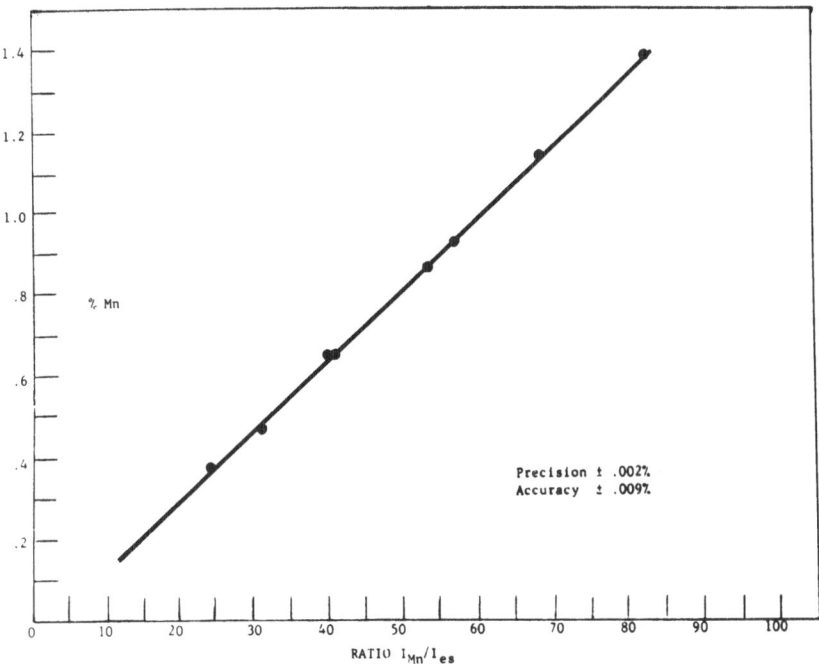

Fig. 16. Mn in cast iron.

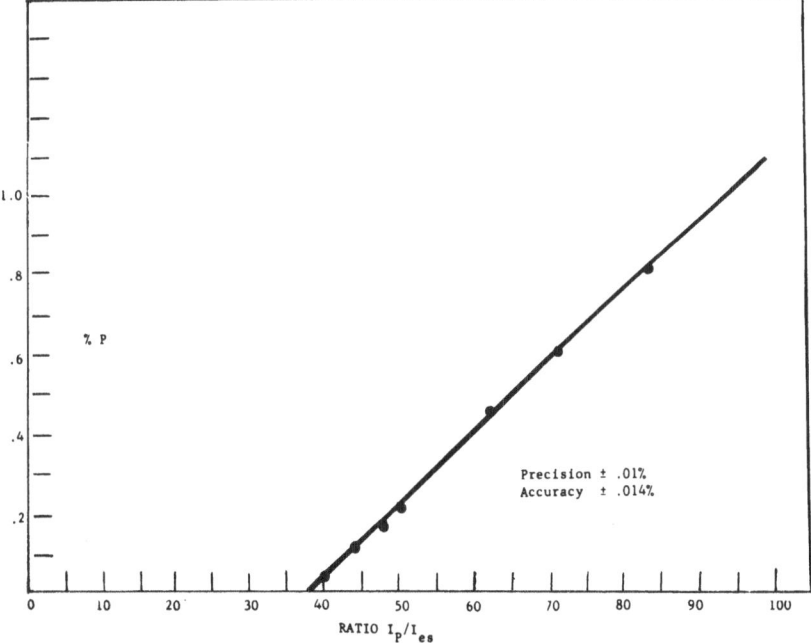

Fig. 17. P in cast iron corrected for high copper.

state. However, accurate control of these materials is not essential and the accuracy shown in Figs. 10–14 is adequate.

The reduction process in the furnace also involves the control of the slag covering the melt since the iron yield is directly dependent upon proper slag composition. The combination of the slag compounds is not always arranged in a reproducible system and, thus, a fusion technique for accurate slag control is sometimes indicated. The analysis of ores and sinters does not involve a matter of the ultimate in speed. However, the proper control of blast-furnace slag requires a determination of the major constituents with a time from sample taking to reporting of results of about 10–15 min.

Our Methods Development Laboratories in Glendale have been studying rapid minimum-flux fusion techniques and have developed one which is fairly simple and straightforward for X-ray fluorescence analyses. Fortunately, slags are easy to fuse and a 1 : 1 sample-to-flux ratio is employed.

The end product of all this effort is the blast-furnace iron or pig iron. Since it is the Mn and Si content, rather than the carbon content, which is of vital interest in the material, X-ray fluorescence techniques offer an area of control for blast-furnace operations.

To summarize, we have presented data to show that multichannel Vacuum X-ray Quantometers provide effective instrumentation for handling another of the important areas of control for the steel industry.

# The Application of X-ray Fluorescence to the Analysis of Highly Alloyed Aluminum Samples

A. C. Ottolini

Research Laboratories
General Motors Corporation

This paper describes a rapid and accurate method for the determination of manganese, copper, iron, nickel, zinc, titanium, and chromium in aluminum alloys by using an X-ray fluorescence technique. The sensitivity of the method allows for the determination of as little as 0.01% of these elements, with the practical concentration ranges being of the order of 0.1 to 3.0%.

A comprehensive program was carried out to measure the precision of the method over a ten-day period. The coefficient of variation was found to be 1.5% for most elements at the 67% confidence level.

Analytical data have been obtained with a view toward employing the same technique for the rapid determination of silicon in the concentration range of 5 to 25%. However, certain variables found in this determination make one proceed with caution, especially when working with highly alloyed material. These variables and their effects will be discussed in detail. Limited data obtained for the determination of magnesium in these alloys will also be reviewed.

Within recent years the metallurgy of light-element alloys has enjoyed an ever-increasing prominence and importance. Throughout the automotive industry and among their suppliers, ambitious research and development programs have been directed toward the field of highly alloyed aluminum. That such an alloy can be feasible for engine construction has been reported by investigators[1,2,3]. The initial dividend of this program at General Motors has been received through the use of an aluminum engine in the Corvair automobile.

However, in the interim and on a continuing basis this program has presented certain challenges to the analytical chemists that have proven quite difficult to overcome. These difficulties are especially evident when applying the X-ray fluorescence technique to the analysis of these alloys.

This paper will discuss our experimental results and conclusions in three phases: first, a discussion detailing the difficulty of determining silicon in these alloys; second, the presentation of test data to demonstrate the feasibility of this technique for alloying elements other than silicon and magnesium; and last, the detection limits for magnesium.

All data reported in this study were obtained using standard Norelco instrumentation, which was selected to meet the needs of our varied workload. The X-ray generator and control, providing up to 50 kv and 50 ma, is standard, except that it is equipped with two high-voltage leads to permit operation of two spectrographic units from the same source.

One spectrograph is equipped with a bulk specimen chamber that allows for excitation of samples as large as 12 in. by 12 in. by 3 in. The specimen in the bulk unit is placed on top of an aluminum mask and excited from the underside. The secondary radiation passes through the collimator to the crystal, where specific wavelengths are diffracted to the counter according to the Bragg equation. This bulk unit is used almost exclusively for elements with atomic numbers greater than 22 (air work). Consequently, it is normally equipped with fine-resolution collimators (parallel plate spacings of 0.005 in.), a lithium fluoride crystal, and a scintillation counter.

The second spectrograph consists of a three-position sample holder equipped with a spinner for rotating samples under excitation, an ethylenediamine ditartrate crystal (EDDT), and a flow-proportional counter, all of which are enclosed in a helium atmosphere. This unit is thus reserved for the determination of elements of atomic number 12 to 22. We consider it quite advantageous to have units available specifically for air work and helium work because we eliminate the need for recalibration and realignment whenever the work dictates a change in counters and crystals.

All secondary radiation measured by the counters is amplified and fed to the electronic circuit panel, where the pulses are counted using a fixed count principle, *i.e.*, a desired fixed level of counts is set into the circuit panel and we measure the time it takes to reach those counts. A simple division gives energy in counts per second. The circuit panel is equipped with a pulse-height analyzer which can

**TABLE I**

Scope

| Element | Concentration range, % |
|---------|------------------------|
| Silicon | 7.0 –25.0 |
| Copper | 0.05–10.0 |
| Manganese | 0.05– 3.0 |
| Nickel | 0.05– 3.0 |
| Iron | 0.05– 2.0 |
| Zinc | 0.05–10.0 |
| Titanium | 0.05– 1.0 |
| Chromium | 0.05– 1.0 |

reduce the background due to scattered radiation and thus increase the signal-to-noise ratio. The application of this unit for discriminating energies occurring from different levels finds considerable application in measuring light-element radiation of the first order which is being interfered with by a higher-order energy from the matrix element. During our aluminum studies, data on silicon and magnesium were obtained using the three-position unit in helium, while data on the other elements were obtained using the bulk unit in air.

As usual for any instrumental work, the major obstacle to method development was in the realm of standards availability. The concentration ranges we desired to cover in these highly alloyed materials are shown in Table I.

Very few commercial standards are available to cover these ranges. Consequently, we embarked on a secondary standards program and our metallurgy group prepared a series of six standards encompassing the range of interest. These standards were prepared under conditions simulating the history and matrix of the unknowns. The standards were certified and checked for the absence of gross lineal inhomogeneity.

Although development was simultaneous for all the elements, let us consider first the many difficulties that arise when one attempts to determine silicon in these alloys. Table II lists the experimental conditions under which the silicon data were obtained.

These conditions give the highest controlled excitation from the primary X-ray tube and thus give the greatest sensitivity in counts per second for silicon. At the same time, the PHA controls the background to the negligible level of 3 to 4 counts/sec by rejecting all

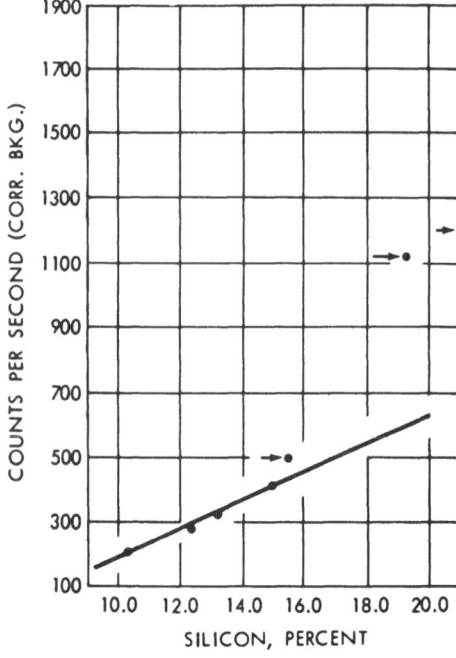

Fig. 1. Silicon in aluminum.

energy below 13 v and above 24 v. Based on the differential curve, these PHA settings yield essentially 90% of the available silicon energy.

Figure 1 shows the analytical curve obtained using the GMR standards and the experimental conditions outlined in Table II. This

### TABLE II
### Operating Conditions for Silicon

| | |
|---|---|
| Spectrograph | Three-position unit |
| Voltage, kv | 50 |
| Amperage, ma | 50 |
| Specimen mask, in. diameter | $1^1/_4$ |
| Atmosphere | Helium |
| Counter | Flow proportional |
| Counter voltage | 1450 |
| PHA baseline, v | 13 |
| PHA window, v | 10.8 |
| Silicon $K_\alpha$, deg $2\theta$ | 108.10 EDDT |
| Background, deg $2\theta$ | 109.50 EDDT |
| Total counts | 16,000 |

figure illustrates the beginning of the downfall of the silicon method using this technique. However, fortunately, or maybe unfortunately, we did not realize it at the time and continued further investigations. The immediate problem is apparent: Does the curve form a straight line as the first few points indicate, or does it bend as much as in-

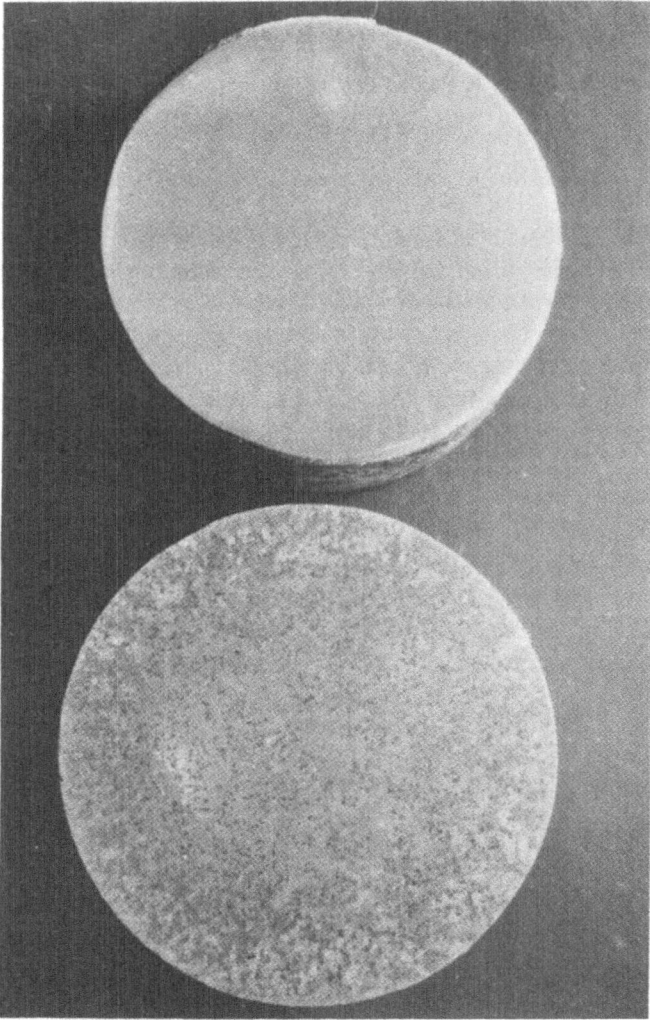

Fig. 2. Mottling. Top, 10% silicon alloy; bottom, GMR-1 standard containing 21% silicon.

dicated by the points marked with an arrow? In addition, there was correlated visual evidence that was quite disturbing, as illustrated in Fig. 2. The shiny, clear-surface sample at the top is a 10% silicon alloy prepared and polished in the same way as its neighbor, which is the GMR X-1 standard containing 21% silicon. The black-speckled or mottled areas are known from metallurgical studies to be crystallized forms of elemental silicon, or magnesium silicides and iron silicides. This mottled condition results when the silicon content in aluminum begins to exceed about 15%, since the eutectic point for true silicon solution in aluminum is roughly at the 12% level. Casting conditions, such as chill rate and phosphorus modification, can begin to control this mottling to a certain extent, but cannot standardize it or eliminate it completely. Since this apparent high counting rate for silicon appeared to increase proportionately with increase of mottling, we were convinced that we were up against a real sampling problem. Microscopic examinations of the finished surfaces substantiated this conclusion because one can see these harder silicides projecting in relief from the softer aluminum surrounding them. From theory we know that the highest percentage of fluorescence radiation coming from the sample occurs at the surface of the specimen; thus, we are convinced that a false indication of silicon energy is obtained because of this mottling.

To completely verify this relationship, we had two series of standards prepared at the 12, 14, 16, 18, and 20% levels of silicon. One series was rapidly chill-cast and phosphorus-modified to reduce the mottling as much as possible, and the other was sand-cast to induce mottling at even the 12% level. Figure 3 shows the result of the study. For each series one gets a fairly smooth and controllable curve for silicon. However, it is evident that the count rate in general is much higher where mottling is high—in fact, by a factor of 50 to 80%. Of course, the immediate question that arises from examining these data is whether one can analyze these samples for silicon using X-ray fluorescence if the metallurgical process is controlled? The data in Fig. 3 marked with crosses partly answer this question. These crosses represent samples prepared under production control at one of our divisions. The scattering of the points between the chill-cast curve and the sand-cast curve makes one wonder whether good reliable answers can be obtained even when the metallurgical process is controlled—at least one should approach such a determination with extreme caution. Obviously, the determination of silicon in fabricated

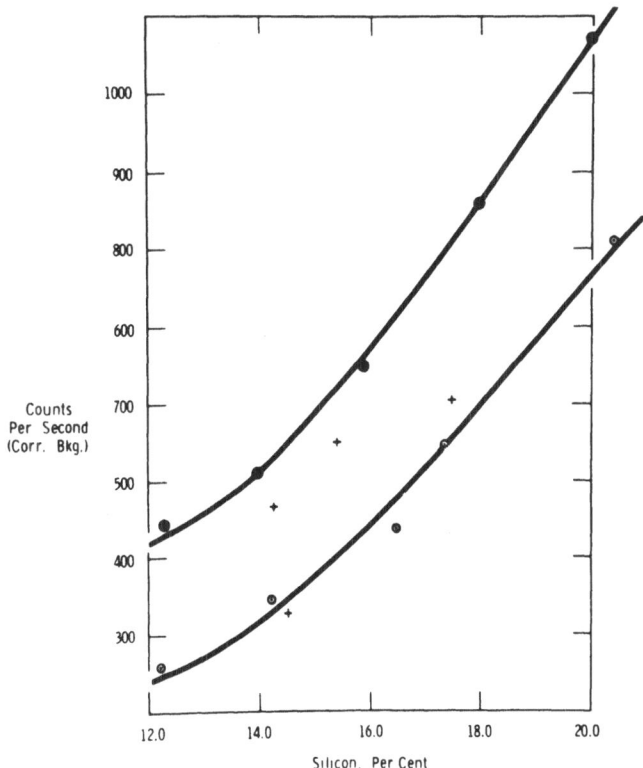

Fig. 3. Sand-cast *vs* chill-cast silicon.

parts and materials from outside vendors must be considered fraught with danger.

At this stage of method development for silicon, two alternatives presented themselves for consideration: first, studying different surface preparations to overcome this mottled effect; second, diluting the samples with an equal amount of pure aluminum to ensure that silicon is below the eutectic point.

We had very little success with our attempts to improve surface finish. Different lathe finishes gave very good repeatable results, but no correction for the mottling effect; the high metallurgical polish finish used for metallographic studies gave identical results with lathe finishes; belt sanding smeared the softer aluminum and thus gave poor results; and finally, both acid and basic etching treatments of the surfaces intensified the silicon count by preferentially

## TABLE III
### Effect of Different Surface Preparations

| Sample | Finish | Counts/sec (corr.) | Si, % |
|--------|--------|--------------------|-------|
| 834 | Lathe (1) | 172.5 | 11.14 |
| 835 | Lathe (1) | 224.0 | 11.21 |
| 834 | Lathe (2) | 169.0 | |
| 835 | Lathe (2) | 220.0 | |
| 834 | M.P. | 174.0 | |
| 835 | M.P. | 227.0 | |
| 834 | L + etch (B) | 294 | |
| 835 | L + etch (B) | 379 | |
| 834 | L + etch | 388 | |
| 835 | L + etch | 497 | |

attacking the more soluble aluminum and leaving more silicon in relief. The only good move we made regarding surface preparation was to send selected specimens to the National Bureau of Standards for study. The results of their specialized surface preparation gave a general improvement in linearity. Because of lack of this equipment, however, we are not in a position to evaluate this surface preparation for all samples and under controlled conditions.

Before undertaking the laborious process of diluting and evaluating these high-silicon alloys for our second approach, we decided to verify our thought that we could determine silicon in the expected diluted range of 5 to 12%. We first secured a group of a dozen aluminum alloys ranging from 6.0 to 11.5%, the chemistry of which had been well established. Each had good, clear surfaces with apparent true solution of silicon throughout the alloy. X-ray examination of these samples indicated that even out of these few selected specimens there were two pairs of samples at the 8 and 11% levels giving results at variance to the level of 30%. Investigation revealed that these samples had essentially the same over-all composition, the same metallurgical history and modification, and the same apparent surfaces. Table III shows the results obtained when working with one pair of these low-silicon alloys. Incidentally, the chemistry of these samples has been verified beyond a doubt.

Metallurgical studies of these samples did reveal, however, that in one sample (835) there was essentially total solution of silicon in

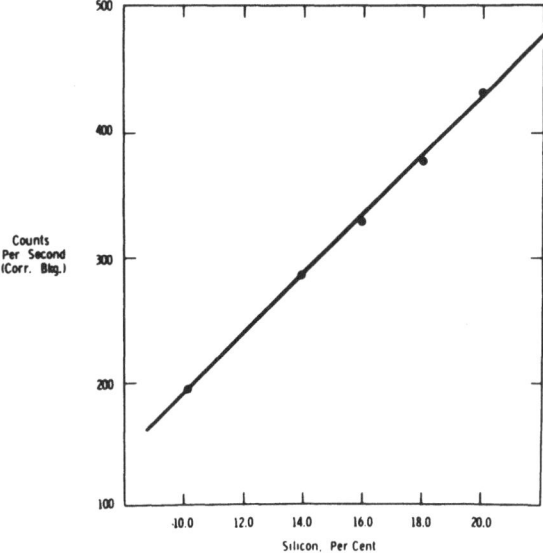

Fig. 4. Silicon in aluminum (powders).

the aluminum, while in the other there was a heavy concentration of silicon present as the ferrosilicide. From theory it is known that iron can have a decided absorption effect on silicon energy and we feel that this is the explanation for the discrepancy in counts per second. However, since the iron content is essentially the same for both samples, how does one control this matrix effect when handling unknown samples on a routine basis? We frankly cannot answer this question, and with the development of a good direct-reading method for this element our interest in this determination is academic at this time.

Lest someone obtains the impression that silicon cannot be determined using X-ray fluorescence, I can report that we control silicon in cast iron at an accuracy of 2% and an elapsed time per determination of 5 min. In addition, with a little less precision, we determine silicon in steels to a concentration level of 0.1%. To evaluate our ability to determine silicon in the presence of aluminum, a series of carefully prepared powdered standards covering a range of 10 to 20% silicon was examined. The results are shown in Fig. 4.

Elements other than silicon and magnesium are easily determined using X-ray fluorescence. Because of the accuracy attainable and the speed of the analysis, this technique serves as our routine

**TABLE IV**
**Operating Conditions (Air)**

| Spectrograph | Bulk unit |
|---|---|
| Voltage, kv | 50 |
| Amperage, ma | 40 |
| Specimen mask, in. diameter | 13/16 |
| Counter | Scintillation |
| Counter voltage, v | 900 |
| PHA baseline, v | 12 |
| Analytical lines | $K_\alpha$ (1st order) |
| Total count | 64,000 to 256,000 |

method for these elements. The samples are machined to a flat, clean surface using a tungsten carbide tool and excited under the conditions listed in Table IV.

The analytical curves obtained for these elements are essentially straight line functions over the concentration ranges of interest. Sensitivity is no problem and for most of these elements the detectable limit is of the order of 0.01%. To ascertain matrix effects due to silicon, the analytical curves have been constructed using two sets of standards, one set containing 4.0–5.0% silicon and the other containing more than 15% silicon. Two typical analytical curves are shown in Figs. 5 and 6. The copper curve has been examined to the 10% level and has been found to bend a little at the higher concentrations. This type of curve structure is consistent with theory for a heavy element in a light matrix. Zinc, iron, nickel, chromium, titanium, and lead are also determined using this technique.

To ascertain the degree of method reproducibility, a test program was carried out over a ten-day period, using two aluminum samples representing high and low concentrations of the elements of interest. Each day the position of the analytical curve was verified using high, middle, and low standards, and the two samples were analyzed as routine unknowns. The results of this test are shown in Table V.

To determine whether there was an effect on these elements due to the degree of mottling on the surface, we selected certain specimens from our sand-cast and chill-cast study which represented extremes of mottling. The results of this study for Mn and Cu are shown in Table VI.

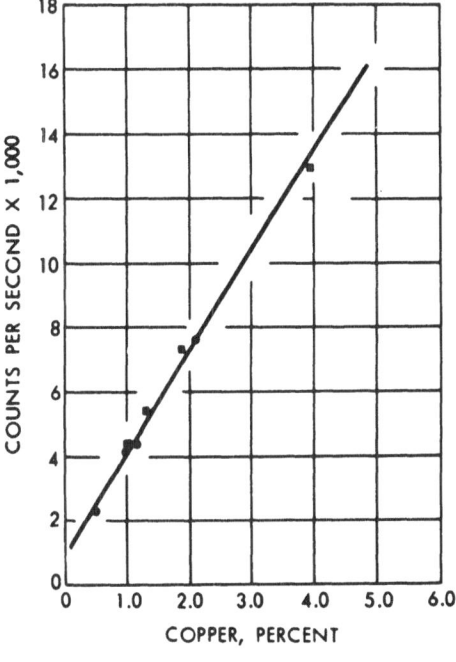

Fig. 5. Copper in aluminum.

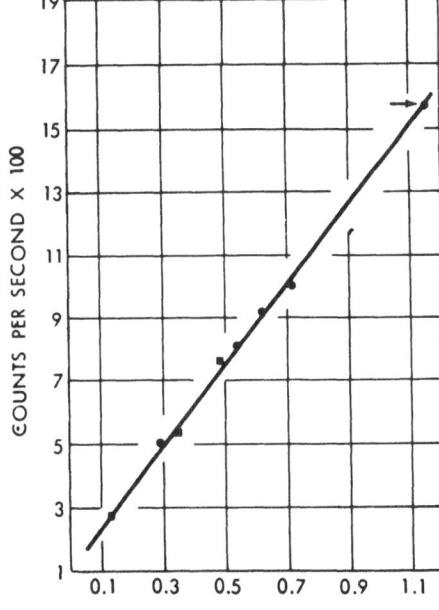

Fig. 6. Manganese in aluminum.

## TABLE V
### Precision Data

| Element | Average concentration, % | n | Coefficient of variation, % |
|---------|--------------------------|---|------------------------------|
| Copper | 1.77 | 10 | 0.95 |
| Copper | 0.17 | 10 | 3.71 |
| Zinc | 1.20 | 10 | 0.48 |
| Zinc | 0.18 | 10 | 1.62 |
| Manganese | 0.51 | 10 | 1.50 |
| Manganese | 0.32 | 10 | 1.96 |
| Nickel | 0.24 | 10 | 1.60 |
| Nickel | 0.11 | 10 | 2.75 |
| Iron | 0.52 | 10 | 2.01 |
| Iron | 0.17 | 10 | 2.24 |

Chemical checks for the other elements confirm the absence of significant effects due to mottling on the accuracy of the method. Since most of the variables were included in the precision study, we feel that the accuracy of the method is almost equivalent to its precision. Chemical checks are made on routine samples selected at random for manganese, iron, and copper, and these data confirm this observation.

Using the three-position unit in helium, we examined the determination of magnesium in these alloys. I can only emphasize at this point that the alignment of the instrumental optics is quite critical for magnesium, since one does not have any intensity to spare. Figure 7 illustrates the analytical curve for magnesium. It is evident that under routine conditions the minimum detectable limit is about

## TABLE VI
### Effect of Mottling

| Sample | Si, % | Mn (x), % | Mn (c), % | Cu (x), % | Cu (c), % |
|--------|-------|-----------|-----------|-----------|-----------|
| 4692 (cc) | 12.0 | 0.57 | 0.59 | 0.97 | 1.01 |
| 4695 (cc) | 16.0 | 0.55 | 0.56 | 1.05 | 1.00 |
| 4695 (sc) | 16.0 | 0.52 | 0.54 | 1.04 | 1.05 |
| 4698 (sc) | 20.0 | 0.53 | 0.55 | 1.03 | 1.07 |

cc = low mottling.
sc = high mottling.

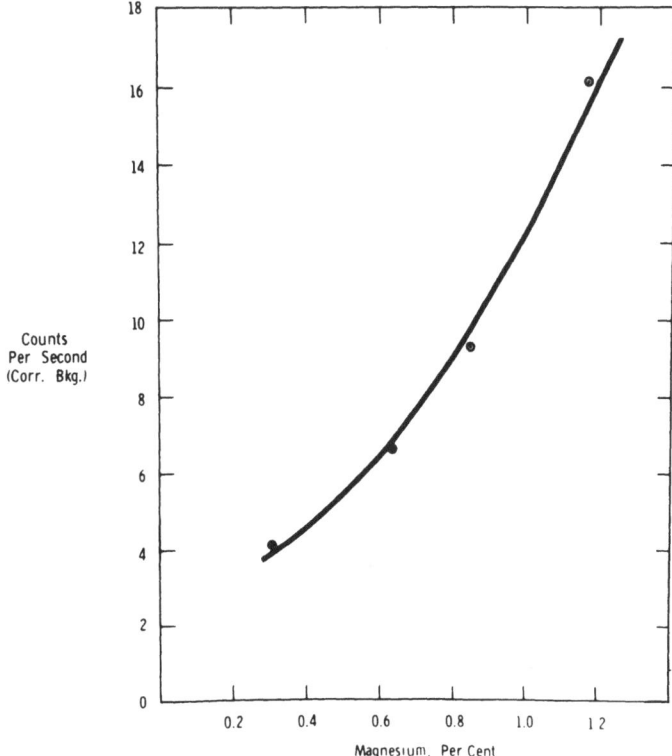

Fig. 7. Magnesium in aluminum.

0.3%. Magnesium was not studied extensively, but I am certain that mottling will have some effect on the results.

To summarize, this has not been the type of paper we are accustomed to presenting to a conference such as this. Usually we like to report an analytical method that has all elements neatly determined and all experimental facets completely controlled. However, we feel that our experiences with silicon in this alloy will prove of value for other investigators and serve as a warning to those who attempt this determination by X-ray fluorescence. For elements other than silicon and magnesium, the X-ray fluorescence method proves to be rapid, precise, and, above all, quite accurate.

## REFERENCES

1.  E. E. Stonebrook, "Aluminum Alloy with High Silicon Content," Talk presented at AFS Convention, May 12, 1960.

2.   K. Schneider, "Aluminum Casting Alloys Improvement by Grain Refinement,"
     Talk presented at AFS Convention, May 12, 1960.
3.   D. F. Caris and R. F. Thomson, "The Implication of the Aluminum Automobile
     Engine," Talk presented at Gray Iron Founders Society Meeting, October 9, 1958.

# Determination of Cladding Thickness of Nuclear Fuel Elements by X-rays

P. Lublin

General Telephone and Electronics Laboratory
Bayside, New York

Cladding on reactor fuel elements is necessary in order to prevent reaction of the uranium core with the surrounding coolant. It may be necessary to keep the cladding between certain minimum and maximum values. Previous methods for determining cladding thickness included eddy-current and radiographic techniques. These methods sometimes fail and an X-ray technique was developed for these cases.

By means of a combined X-ray spectrographic–absorption technique, similar to that used in the tin-plating industry, cladding thickness can be determined accurately. The application of this technique to zirconium and aluminum cladding on uranium-core material will be discussed. Mention will be made of other materials examined by other X-ray techniques.

# Some Aspects of X-ray Spectrography of Thin Films

J. Sherman

Philadelphia Naval Ship Yard
Philadelphia, Pennsylvania

Certain peculiar and even anomalous variations in the X-ray emission from thin films may be encountered. A theoretical and experimental study of some of these efforts is presented.

# A Study of the Complete Pattern of Curves for a Variety of Matrices

## B. J. Mitchell

Union Carbide Metals Co.
Niagara Falls, New York

The variation in the fluorescent intensity of a given-wavelength X-radiation with its matrix has been found to be generally inverse to the variation in its mass absorption coefficient. Intensity–matrix correlations have been studied for iron, chromium, thorium, manganese, calcium, silicon, tantalum, columbium, titanium, and zirconium, and complete systems of calibration curves developed therefrom. A simple method for representing a family of curves based on the intensity of the pure material is described.

# The Analysis of 15–W, 5–Mo, 1–Zr, Cb Alloy by X-ray Fluorescence Spectroscopy

### E. E. Hanson

Wah Chang Corp.
Albany, Oregon

The analysis of the F-48 Alloy illustrates the great versatility of the X-ray spectrograph as an analyst's tool. The average instrument can be used for the analysis of the alloying constituents as well as the tantalum impurity. This wide range extends from the 15% level for tungsten to 100 ppm for tantalum.

Certain resolution and intensity problems are encountered which can be resolved by the use of fine collimation and a topaz crystal; extra intensity for tantalum is gained by the use of a gold target tube.

The required sensitivity for tantalum demands the use of an oxide sample; this is prepared by fuming the dissolved sample with perchloric acid and using a controlled burning temperature to prohibit the loss of molybdenum.

# Fast–Scanning Method for X-ray Spectroscopy

M. L. Salmon

Fluo-X-Spec Laboratories
Denver, Colorado

Commercial instrumentation usually provides a range of goniometer scanning rates for selection by the operator. Succcessful application of the faster scanning rates is dependent on several features of the instrument in addition to the precise mechanical movement of the goniometer. Important features are the response rates of scaler, ratemeter, and recorder circuits and mechanical motion characteristics of recorder indicating devices.

Satisfactory results are evaluated on the basis of precision and accuracy of the indicated intensity values as a function of $2\theta$ angle values on the chart recording, with special emphasis on the degree of resolution indicated in the chart trace.

It is sometimes desirable to scan at rates faster than those normally provided in standard commercial instrumentation, and slight modifications of a standard 100–kv Norelco spectrograph will be outlined in a discussion of a study of scanning rates as fast as 32 degrees $2\theta$ per minute.

# Methods to Increase X-ray Sensitivity

W. R. Kiley

Philips Electronics
Mount Vernon, New York

The X-ray spectrograph basically consists of three components that determine the sensitivity:

1. Source of excitation, which includes the X-ray tube target and the type of high voltage generator.

2. Optics, which includes collimation and the choice of crystal monochromator.

3. Detectors, including pulse-height selection.

Comparison data shown for most of the commercially available components, and intensities for available samples of light elements are given.

# On-Line Process Analysis by X-ray Emission Techniques

W. F. Loranger

General Electric Co.
Milwaukee, Wisconsin

Instrumentation designed to utilize the principles of X-ray emission optics will produce an output signal proportional to the amount of an element present in a material, whether the material is *stationary* or *moving*. Such instrumentation, therefore, is very well suited for continuous analysis of elements in a material flowing in a process line.

The General Electric X-Ray Emission Gage (XEG) was designed for and is being used as such an on-line chemical sensor. It is capable of measuring up to six elements in a material simultaneously. The range of elements is generally from uranium to aluminum, and the material analyzed may be in a variety of states (liquid, slurry, powder, etc.). The output signals may be calibrated to read percent, weight, thickness, etc., and are suitable for use with recorders, data loggers, computers, and/or for direct control devices.

The instrumentation will be described and calibration data for several elements in a variety of materials will be shown. (All data taken were obtained on moving samples.)

# Ultraviolet and
# Visible Spectroscopy

# Quality Control of Steel Using Clock- and Chart-Recording Photoelectric Spectrometers*

Joseph F. Woodruff and Arba H. Thomas

Chemical Research Department
Research and Technology, Armco Steel Corporation
Middletown, Ohio

A check-analysis program initiated at Armco over 25 years ago is used to evaluate the performance of two ARL Quantometers and three Baird-Atomic Direct Readers in five analytical control laboratories of the Armco Steel Corporation. The program has also been used to evaluate wet chemical analyses reported by these same laboratories prior to the installation of photoelectric spectrometers.

The paper describes the check-analysis program and analytical data used to evaluate the accuracy of emission photoelectric spectrometers.

## INTRODUCTION

Since the early 1950s, production metallurgists and chemists of the Armco Steel Corporation have installed emission photoelectric spectrometers to replace the chemical procedures for production control. These instruments have been used for control of melts produced in the cupola and in the blast, open-hearth, and electric furnaces.

The use of this equipment in production control has resulted in savings of analytical manpower, heat time, and scrapped heats. There has been an increase in the amount of steel produced in a number of

---

*Paper presented at the 12th Annual Symposium on Spectroscopy of the Chicago Section of the Society of Applied Spectroscopy at the Conrad Hilton Hotel, Chicago, Illinois, May 15 to May 18, 1961; the Pittsburgh Conference on Analytical Chemistry and Applied Spectroscopy, Feb. 27 to March 3, 1961, at the Hotel Penn-Sheraton; and the General Motors Spectrographic Conference at the General Motors Technical Center, Warren, Michigan, April 24 to April 25, 1961.

melt shops because of more rapid reporting of the analyses after a preliminary furnace test is received. With more information available, metallurgists have a better understanding of the effects of trace and residual elements, which results in better quality steel. In a matter of minutes, melters and metallurgists are given information which enables them to divert melts and make accurate alloy additions. Prior to installation of photoelectric spectrometers it was not unusual for a melter to wait 30 to 40 minutes for a nickel or chromium determination.

Even with all these advantages of instrumental analysis, the accuracy of the analytical results cannot be sacrificed for speed. Inaccurate analyses cannot be tolerated since losses in production, quality, and profits to the company would follow. Therefore, any photoelectric spectrometer installed in a production laboratory must meet or surpass the accuracy of the chemical methods formerly used.

The purpose of this paper is to show that a photoelectric spectrometer is a precise and rapid analytical tool and can control the chemistry of the melts produced in a steel plant.

## CHECK-ANALYSIS PROGRAM

A check-analysis program [1] was initiated at Armco over 25 years ago to determine the accuracy of the analytical values reported by the production control laboratories on the heats and casts produced. The supervisor in charge of each laboratory makes out a daily report and certifies the ladle analyses as determined by the "turn" analysts on all heats and casts produced for a 24-hr period. The Chemical Research Department reviews these daily reports and selects one day's analytical production for checking. The day selected is varied from month to month. A wire is sent to a particular production laboratory requesting ladle samples of selected heats and casts produced on the selected day.

In selecting the determinations to be checked, an attempt is made to obtain and analyze an equal number from each of the three work shifts so that the work of all analysts for that 24-hr period is checked. So far as is feasible, samples from various grades of steel are selected. As soon as the analyses are completed by Research, the production laboratory is advised by telephone or telegram of those determinations certified on the heat report that are outside the

Armco standard analytical tolerances. A recheck of those analyses that exceed the tolerance units is then made by the production control laboratory, and the recheck values reported to Research. In instances where the recheck values agree with the original certified values, samples are sent to other Armco control laboratories for analysis. If the third laboratory agrees with the production laboratory, the original analyses of the latter are considered correct.

As a basis for determining whether or not a value reported by the Works laboratory is in error, standard analytical tolerances have been established. These tolerances are based on the accuracies obtained when using the best chemical procedures available. Table I shows the Armco allowable tolerances permitted for each element according to the concentration range and the type of steel or iron involved. Only those elements that are normally analyzed by the photoelectric spectrometers are included in the table. The value reported by the production laboratory is in error if it is not in agreement with the referee Research value or the referee value of a third laboratory, based on the tolerances shown in Table I.

The concluding step in the check-analysis program is the issuance of a report showing the Research and production values for all the determinations checked for that particular month. A typical page of one of these reports is shown in Fig. 1. Each laboratory is given a percentage rating based on a credit of 100 for each determination within tolerance and a credit of zero for those determinations in which the differences are two or more times the standard tolerance limits. Partial credit is given for those differences which fall between one and two times the standard tolerance limits. The sum of all the credits, divided by the number of determinations, gives the percentage rating. An example of how the percentage rating is calculated is shown in Table II.

As can be seen from this description of the check-analysis program, we have a basis for determining whether or not the photoelectric spectrometers give reliable analyses equal to or better than those given by the chemical methods formerly used. Using the reports from the check-analysis program, a study was made of the accuracy of the photoelectric spectrometers since their installation. A similar study was made of the accuracy of analyses for a five-year period prior to the use of photoelectric spectrometers, when only wet chemical methods were used.

## TABLE I
### Standard Analytical Tolerances for Armco Analytical Laboratories

| Type of steel | Elements | Concentration range, % element | Tolerance, plus or minus % element |
|---|---|---|---|
| Carbon and low alloy | Mn | 0.000– 0.099 | 0.003 |
| | | 0.10 – 0.39 | 0.01 |
| | | 0.40 – 0.99 | 0.02 |
| | | 1.00 – 1.99 | 0.03 |
| Stainless and high alloy | Mn | 0.000– 0.099 | 0.005 |
| | | 0.10 – 0.49 | 0.02 |
| | | 0.50 – 1.49 | 0.03 |
| | | 1.50 – 2.49 | 0.04 |
| | | 2.50 – 4.99 | 0.06 |
| | | 5.00 – 9.99 | 0.10 |
| | | 10.00 –19.99 | 0.15 |
| Carbon and low alloy | P | 0.000– 0.049 | 0.003 |
| | | 0.050– 0.099 | 0.005 |
| | | 0.10 – 0.29 | 0.01 |
| Stainless and high alloy | P | 0.000– 0.014 | 0.002 |
| | | 0.015– 0.049 | 0.003 |
| | | 0.050– 0.099 | 0.005 |
| | | 0.10 – 0.29 | 0.01 |
| | | 0.30 – 0.49 | 0.02 |
| Carbon and low alloy | Si | 0.000– 0.009 | 0.003 |
| | | 0.010– 0.099 | 0.005 |
| | | 0.10 – 0.29 | 0.01 |
| | | 0.30 – 0.99 | 0.02 |
| | | 1.00 – 2.99 | 0.03 |
| | | 3.00 – 5.99 | 0.04 |
| Stainless and high alloy | Si | 0.000– 0.099 | 0.006 |
| | | 0.10 – 0.39 | 0.01 |
| | | 0.40 – 0.99 | 0.02 |
| | | 1.00 – 2.99 | 0.03 |
| | | 3.00 – 5.99 | 0.04 |
| Carbon and low alloy | Cu | 0.000– 0.099 | 0.006 |
| | | 0.10 – 0.39 | 0.01 |
| | | 0.40 – 0.69 | 0.02 |
| | | 0.70 – 0.99 | 0.03 |
| Stainless and high alloy | Cu | 0.000– 0.099 | 0.006 |
| | | 0.10 – 0.39 | 0.01 |
| | | 0.40 – 0.69 | 0.02 |
| | | 0.70 – 0.99 | 0.03 |
| | | 1.00 – 1.99 | 0.04 |
| | | 2.00 – 2.99 | 0.05 |
| | | 3.00 – 3.99 | 0.06 |

TABLE I (*continued*)

| Type of steel | Elements | Concentration range, % element | Tolerance, plus or minus % element |
|---|---|---|---|
| Carbon and low alloy | Cr, Mo, Ni, Co, Al, Ti, Zr, Sn, As, Sb | 0.000– 0.009<br>0.010– 0.049<br>0.050– 0.099<br>0.10 – 0.29<br>0.30 – 0.99 | 0.002<br>**0.002**<br>0.005<br>0.01<br>0.02 |
| Stainless and high alloy | Cb | **0.000– 0.099**<br>0.10 – 0.29<br>0.30 – 0.49<br>0.50 – 0.99<br>1.00 – 1.99 | 0.01<br>0.02<br>0.03<br>0.04<br>0.06 |
| Stainless and high alloy | Co | 0.000– 0.099<br>0.10 – 0.29<br>0.30 – 0.49<br>0.50 – 0.99<br>1.00 – 1.99 | 0.005<br>0.01<br>0.02<br>0.03<br>0.05 |
| Stainless and high alloy | Al | 0.000– 0.019<br>0.020– 0.099<br>0.10 – 0.15<br>0.16 – 0.49<br>0.50 – 0.79<br>0.80 – 1.49<br>1.50 – 1.99 | 0.003<br>0.005<br>0.01<br>0.02<br>0.04<br>0.05<br>0.06 |
| Stainless and high alloy | Ni, Cr | 0.000– 0.099<br>0.10 – 0.29<br>0.30 – 0.99<br>1.00 – 1.99<br>2.00 – 2.99<br>3.00 – 3.99<br>4.00 – 5.99<br>6.00 – 7.99<br>8.00 – 9.99<br>10.00 –14.99<br>15.00 –29.99<br>30.00 –49.99<br>50.00 and greater | 0.005<br>0.01<br>0.02<br>0.03<br>0.04<br>0.05<br>0.06<br>0.07<br>0.08<br>0.10<br>0.15<br>0.20<br>0.25 |
| Stainless and high alloy | Mo | 0.000– 0.099<br>0.10 – 0.29<br>0.30 – 0.69<br>0.70 – 0.99<br>1.00 – 1.99<br>2.00 – 3.99 | 0.005<br>0.01<br>0.02<br>0.03<br>0.04<br>0.06 |

Sheet No. 6

Date: March, 1960

Works A

| Heat | Grade | Turn | Wks | C | Mn | P | S | Si | Cu | Ni | Cr | Mo | Al |
|---|---|---|---|---|---|---|---|---|---|---|---|---|---|
| 1 | A | 3-4 | Res. | | | | | 3.22 w | | | | | |
| | | | Lab A | | | | | 3.22 w | | | | | |
| 2 | A | 12-8 | Res. | .027 c | | .006 w | .026 c | 3.08 w | | | | | |
| | | | Lab A | .029 c | | .006 w | .027 c | 3.10 q | | | | | |
| 3 | A | 4-12 | Res. | | | | | 3.08 w | | | | | |
| | | | Lab A | | | | | 3.06 q | | | | | |
| 4 | B | 8-4 | Res. | .083 c | .41 w | .006 w | .026 c | | .12 w | | | | |
| | | | Lab A | .087 c | .41 q | .009 w | .027 c | | .11 q | | | | |
| 5 | C | 12-8 | Res. | .67 c | .78 w | | | .30 w | | | | | |
| | | | Lab A | .67 c | .74 w* (.77 w) | | | .29 w | | | | | |
| 6 | C | 4-12 | Res. | | | .009 w | .019 c | .26 w | | | | | |
| | | | Lab A | | | .010 w | .021 c | .27 w | | | | | |
| 7 | D | 12-8 | Res. | | .53 w | | | | .13 w | 7.20 w | 17.06 w | .11 w | 1.18 w |
| | | | Lab A | | .55 q | | | | .14 q | 7.13 q | 17.15 q | .12 q | 1.19 q |
| 8 | E | 8-4 | Res. | | | | | .48 w | .15 w | 9.01 w | 19.03 w | | |
| | | | Lab A | | | | | .50 q | .16 q | 9.04 q | 18.99 q | | |
| 9 | F | 4-12 | Res. | .041 c | 1.82 w | | | | | 13.42 w | 17.37 w | 2.18 w | |
| | | | Lab A | .045 c | 1.85 q | | | | | 13.42 q | 17.44 q | 2.24 q | |
| 10 | G | --- | Res. | | | | | | | 12.24 w | | | |
| | | | Lab A | | | | | | | 12.38 q* (12.31 w) | | | |
| 11 | C | --- | Res. | | | | .015 c | | | | | | 1.15 w |
| | | | Lab A | | | | .016 c | | | | | | 1.17 q |
| 12 | H | --- | Res. | | | | .010 c | | .081 w | .39 w | | | |
| | | | Lab A | | | | .009 c | | .080 q | .37 q | | | |

Ratings :  Chemical 97%   Spectrochemical 98.2%   Works 98.1%

*These values are outside of tolerance limits. Check values in parentheses
(c) combustion (w) wet chemical (q) Quantometer (x) Direct Reader (o) Spectro-Photo

Fig. 1. Typical laboratory check-analysis report.

## TABLE II
### Check-Analysis Grading System—Method
### of Calculating Percentage Rating

| Tolerance, plus or minus % element | Deviation of reported value from research value, % element | Credit allowed, % |
|---|---|---|
| 0.002 | 0.002 | 100 |
| 0.002 | 0.003 | 50 |
| 0.002 | 0.004 | 0 |
| 0.003 | 0.003 | 100 |
| 0.003 | 0.004 | 67 |
| 0.003 | 0.005 | 33 |
| 0.003 | 0.006 | 0 |
| 0.005 | 0.005 | 100 |
| 0.005 | 0.006 | 80 |
| 0.005 | 0.007 | 60 |
| 0.005 | 0.008 | 40 |
| 0.005 | 0.009 | 20 |
| 0.005 | 0.010 | 0 |
| 0.04 | 0.04 | 100 |
| 0.04 | 0.05 | 75 |
| 0.04 | 0.06 | 50 |
| 0.04 | 0.07 | 25 |
| 0.04 | 0.08 | 0 |
| 0.10 | 0.10 | 100 |
| 0.10 | 0.11 | 90 |
| 0.10 | 0.12 | 80 |
| 0.10 | 0.13 | 70 |
| 0.10 | 0.14 | 60 |
| 0.10 | 0.15 | 50 |
| 0.10 | 0.16 | 40 |
| 0.10 | 0.17 | 30 |
| 0.10 | 0.18 | 20 |
| 0.10 | 0.19 | 10 |
| 0.10 | 0.20 | 0 |

Example

| 15 determinations within tolerance limit × 100 | 1500 |
|---|---|
| 2 determinations outside double tolerance limit × 0 | 0 |
| 1 determination (partial credit) × 50 | 50 |
| 1 determination (partial credit) × 70 | 70 |
| 1 determination (partial credit) × 30 | 30 |
| Total credits | 1650 |

(Total credits) (Number of determinations) = Percentage rating

1650/20 = 82.5% rating

## TABLE III
### Accuracy of Photoelectric Spectrometer vs Research Referee Values for Elements Determined in Carbon, Low-Alloy, and Stainless Steels during Monthly Check-Analysis Program

| Element | Spectrometer | Concentration range, % element | Average concentration, % element | Average difference from referee values, % element | Average difference from referee values, % of amount present | Number of values compared |
|---|---|---|---|---|---|---|
| Mn | A | 0.027– 0.099 | 0.085 | 0.003 | 3.5 | 69 |
| Mn | B | 0.017– 0.097 | 0.039 | 0.003 | 7.7 | 52 |
| Mn | A | 0.10 – 0.92 | 0.36 | 0.010 | 2.8 | 227 |
| Mn | B | 0.10 – 0.98 | 0.49 | 0.010 | 2.0 | 459 |
| Mn | A | 1.00 – 2.00 | 1.35 | 0.020 | 1.5 | 67 |
| Mn | B | 1.00 – 2.01 | 1.50 | 0.030 | 2.0 | 122 |
| Si | B | 0.021– 0.094 | 0.054 | 0.004 | 7.4 | 63 |
| Si | A | 0.10 – 0.81 | 0.39 | 0.009 | 2.3 | 269 |
| Si | B | 0.10 – 0.88 | 0.42 | 0.025 | 6.0 | 366 |
| Si | A | 1.09 – 3.72 | 2.66 | 0.030 | 1.1 | 211 |
| Si | B | 1.20 – 5.06 | 3.22 | 0.040 | 1.2 | 83 |
| Cu | A | 0.020– 0.096 | 0.072 | 0.002 | 2.8 | 9 |
| Cu | B | 0.035– 0.099 | 0.066 | 0.003 | 4.5 | 94 |
| Cu | A | 0.10 – 0.54 | 0.18 | 0.009 | 5.0 | 47 |
| Cu | B | 0.10 – 0.80 | 0.23 | 0.009 | 4.0 | 322 |
| Cu | B | 1.02 – 3.69 | 3.18 | 0.03 | 0.9 | 18 |
| Ni | A | 0.047– 0.079 | 0.061 | 0.004 | 6.6 | 13 |
| Ni | B | 0.008– 0.090 | 0.029 | 0.002 | 7.0 | 323 |
| Ni | A | 0.11 – 0.78 | 0.45 | 0.010 | 2.2 | 57 |
| Ni | B | 0.10 – 0.71 | 0.26 | 0.008 | 3.1 | 31 |
| Ni | A | 1.03 –13.32 | 7.64 | 0.045 | 0.6 | 156 |

| | | Range | | | | |
|---|---|---|---|---|---|---|
| Cr | A | 0.013– 0.061 | 0.028 | 0.002 | 7.1 | 13 |
| Cr | B | 0.050– 0.090 | 0.70 | 0.003 | 4.3 | 9 |
| Cr | A | 0.34 – 0.98 | 0.66 | 0.01 | 0.7 | 67 |
| Cr | B | 0.15 – 0.45 | 0.39 | 0.01 | 2.6 | 8 |
| Cr | A | 1.02 –19.04 | 16.62 | 0.08 | 0.5 | 213 |
| Mo | A | 0.007– 0.032 | 0.012 | 0.002 | 30.0 | 20 |
| Mo | B | 0.007– 0.070 | 0.031 | 0.004 | 12.9 | 50 |
| Mo | A | 0.10 – 0.62 | 0.24 | 0.009 | 3.6 | 83 |
| Mo | B | 0.10 – 0.59 | 0.40 | 0.011 | 2.7 | 38 |
| Mo | B | 2.04 – 3.75 | 2.44 | 0.055 | 2.2 | 84 |
| Al | A | 0.004– 0.008 | 0.0052 | 0.0004 | 7.7 | 164 |
| Al | B | 0.003– 0.036 | 0.0044 | 0.0005 | 10.9 | 226 |
| Al | A | 0.23 – 0.45 | 0.34 | 0.010 | 2.9 | 163 |
| Al | B | 0.18 – 0.50 | 0.32 | 0.005 | 1.6 | 66 |
| Al | A | 1.14 – 1.30 | 1.20 | 0.015 | 1.2 | 12 |
| Al | B | 1.00 – 1.36 | 1.18 | 0.040 | 3.4 | 49 |
| Cb | B | 0.28 – 0.95 | 0.57 | 0.015 | 2.6 | 13 |

## TABLE IV
### Accuracy of Photoelectric-Spectrometer and Routine Chemical vs Research Referee Values for Elements Determined in Carbon, Low-Alloy, and Stainless Steels during Monthly Check-Analysis Program

| Element | Routine method | Concentration range, % element | Average concentration, % element | Average difference from referee values, % element | Average difference from referee values, % of amount present | Number of values compared |
|---|---|---|---|---|---|---|
| Mn | Spectrometers | 0.017–0.099 | 0.065 | 0.0030 | 4.6 | 121 |
| Mn | Chemical | 0.009–0.098 | 0.042 | 0.0024 | 5.7 | 110 |
| Mn | Spectrometers | 0.10 –0.98 | 0.45 | 0.010 | 2.2 | 686 |
| Mn | Chemical | 0.10 –0.99 | 0.49 | 0.012 | 2.4 | 718 |
| Mn | Spectrometers | 1.00 –2.01 | 1.44 | 0.026 | 1.8 | 189 |
| Mn | Chemical | 1.00 –9.20 | 1.59 | 0.028 | 1.7 | 220 |
| Si | Spectrometers | 0.021–0.094 | 0.054 | 0.0040 | 7.4 | 63 |
| Si | Chemical | 0.002–0.099 | 0.044 | 0.0035 | 7.9 | 153 |
| Si | Spectrometers | 0.10 –0.88 | 0.41 | 0.018 | 4.4 | 635 |
| Si | Chemical | 0.12 –0.91 | 0.35 | 0.009 | 2.6 | 391 |
| Si | Spectrometers | 1.09 –5.06 | 2.82 | 0.032 | 1.1 | 294 |
| Si | Chemical | 1.27 –5.15 | 2.95 | 0.023 | 0.8 | 130 |
| Cu | Spectrometers | 0.020–0.099 | 0.066 | 0.0030 | 4.5 | 103 |
| Cu | Chemical | 0.010–0.098 | 0.060 | 0.0020 | 3.3 | 129 |
| Cu | Spectrometers | 0.10 –0.80 | 0.22 | 0.009 | 4.1 | 369 |
| Cu | Chemical | 0.10 –0.60 | 0.23 | 0.007 | 3.0 | 201 |
| Ni | Spectrometers | 0.008–0.090 | 0.030 | 0.0021 | 7.0 | 336 |
| Ni | Chemical | 0.040–0.098 | 0.068 | 0.0064 | 9.4 | 29 |
| Ni | Spectrometers | 0.10 –0.78 | 0.37 | 0.009 | 2.4 | 88 |
| Ni | Chemical | 0.10 –0.78 | 0.32 | 0.016 | 5.0 | 125 |

| | | | | | | |
|---|---|---|---|---|---|---|
| Ni | Spectrometers | 1.03 –13.32 | 7.64 | 0.045 | 0.6 | 156 |
| Ni | Chemical | 1.13 –19.68 | 9.72 | 0.072 | 0.7 | 315 |
| Cr | Spectrometers | 0.013– 0.090 | 0.045 | 0.0020 | 4.4 | 22 |
| Cr | Chemical | 0.015– 0.090 | 0.054 | 0.0040 | 7.4 | 29 |
| Cr | Spectrometers | 0.15 – 0.98 | 0.63 | 0.010 | 1.6 | 75 |
| Cr | Chemical | 0.10 – 0.43 | 0.31 | 0.008 | 2.6 | 39 |
| Cr | Spectrometers | 1.02 –19.04 | 16.62 | 0.08 | 0.5 | 213 |
| Cr | Chemical | 5.41 –26.00 | 17.00 | 0.06 | 0.4 | 425 |
| Mo | Spectrometers | 0.007– 0.070 | 0.026 | 0.0030 | 11.5 | 70 |
| Mo | Chemical | 0.004– 0.095 | 0.045 | 0.0080 | 17.8 | 76 |
| Mo | Spectrometers | 0.10 – 0.62 | 0.29 | 0.010 | 3.4 | 121 |
| Mo | Chemical | 0.10 – 0.62 | 0.31 | 0.013 | 4.2 | 191 |
| Mo | Spectrometers | 2.04 – 3.75 | 2.44 | 0.055 | 2.2 | 84 |
| Mo | Chemical | 2.01 – 3.20 | 2.36 | 0.049 | 2.1 | 53 |
| Cb | Spectrometers | 0.13 – 0.87 | 0.50 | 0.030 | 6.0 | 53 |
| Cb | Chemical | 0.28 – 0.95 | 0.57 | 0.015 | 2.6 | 13 |

## Accuracy of Type A vs Type B Photoelectric Spectrometers

Two different types of emission photoelectric spectrometers have installed in the production laboratories. For obvious reasons these instrum ˑʌts will not be identified by name in the discussions to follow.

The accuracy of the two types of photoelectric spectrometers, based on data from the reports of the monthly check-analysis program, is shown in Table III.

In general, the accuracy of analysis is approximately the same for instruments A and B for the elements in the concentration ranges given in Table III. However, type A instruments give more accurate results than type B for silicon determinations between 0.10 and 1.00% silicon, nickel determinations below 0.10%, molybdenum determinations between 0.10 and 0.60%, and aluminum determinations between 1.00 and 1.40%. Instruments of type B give more accurate results than type A for aluminum determinations in the range of 0.10 to 0.50%.

Differences in sampling techniques, source conditions, personnel, *etc.*, could contribute to the variation in accuracy of these photoelectric spectrometers. The spectrometric methods are not standardized and each laboratory has set up its own procedures.

Overall, the average analytical error for elements in concentrations greater than 1.0% is approximately 1.0% when using instruments A and 2.0% when using instrument B. The average analytical error for elements in the concentration range of 0.1 to 0.99% is approximately 3.0% when using either of the two instruments. For elements below 0.1% in concentration, the average error is approximately 9.0% for the two instruments. The error for some elements is much less than for others, as can be seen from Table III, and depends largely on the mean concentration of these elements.

In the foregoing statements we are assuming that the Research referee values are accurate and without error. However, the average differences between the spectrometer and Research referee values include the errors in both methods.

## Accuracy of Production-Control Chemical Methods vs Accuracy of Spectrometric Methods

To further confirm the reliability of the emission-photoelectric-spectrometer methods, a second study was made. A comparison of routine chemical and photoelectric-spectrometer values was made with Research referee values as determined by the monthly check-analysis program. These data are shown in Table IV. Data for the routine chemical analyses were obtained from the check-analysis reports for the period of five years prior to the installation of photo-electric spectrometers in the five production control laboratories.

With few exceptions, the values obtained by the photoelectric spectrometers met the same analytical tolerances used for chemical methods. The analytical errors of the routine chemical methods were not as great as the errors of the spectrometric methods for the determinations of silicon in the range of 0.01 to 0.91% and columbium in the range of 0.13 to 0.95%. However, the spectrometric methods gave more accurate results for the nickel, chromium, and molybdenum determinations in concentrations below 1.0%.

From this study it is evident that the production control laboratories did not sacrifice accuracy for speed when they installed spectrometers. In fact, they improved the accuracy for the determination of nickel, chromium, and molybdenum in concentrations below 1.0%.

## DISCUSSION AND CONCLUSIONS

A total of 4336 spectrometer analyses from five different production laboratories of the Armco Steel Corporation have been checked by referee chemical methods in a check-analysis program established at Armco over 25 years ago. Analyses from these same five laboratories, when routine chemical methods were used, were also evaluated by the check-analysis system and used in this study for comparison with the photoelectric-spectrometer data.

Evaluation of the data for these two methods of analysis show that 88% of the routine chemical and 87% of the photoelectric spectrometer analyses compared were within the rigid analytical tolerances established for the Armco check-analysis program. If we consider only the past two years, the operating laboratories have had 94% of the spectrometer analyses within the analytical tolerances.

These studies show that instrumental methods can replace the conventional chemical methods without loss of accuracy in production control laboratories.

## ACKNOWLEDGMENT

We would like to thank the Armco Steel Corporation for their permission to publish this paper.

## REFERENCE

1.  Arba H. Thomas and Charles S. Mills, "Interplant Standardization in the Steel Industry," *Anal. Chem.* **23**, 1553–1555, 1951.

# Establishing and Controlling Analytical Curves

## T. P. Schreiber

Research Laboratories
General Motors Corporation
Warren, Michigan

The accuracy of many analytical methods is determined to a large extent by the accuracy and day-to-day reliability of an analytical curve.

The initial establishment of an analytical curve involves the random errors associated with repeatability and systematic errors which may be inherent in the method or the standards used for calibration. When the calibration data result in randomly scattered points, a simplified method of least squares can be used to draw the most reliable curve.

Daily control of the analytical curve is a definite problem, as indicated by data from a group of spectrochemical laboratories. Statistical techniques can be used to evaluate and correct for these daily variations.

## INTRODUCTION

An analytical curve is a deceptively simple device. As indicated in Fig. 1, it is merely a line which relates a characteristic which can be measured to the characteristic which must be determined. In emission spectroscopy these characteristics are intensity ratio and element concentration, respectively. For a particular experimental arrangement and analysis, e.g., Mn in steel, there exists a "true" analytical curve which accurately reflects the relationship between log % Mn and the log intensity ratio of a Mn line to an Fe line. If the analyst runs standard samples and, because of the statistical variations in the intensity ratio, obtains an analytical curve different from the true curve, then the % Mn values obtained from this curve will have systematic errors in many areas. Although statistical variations can

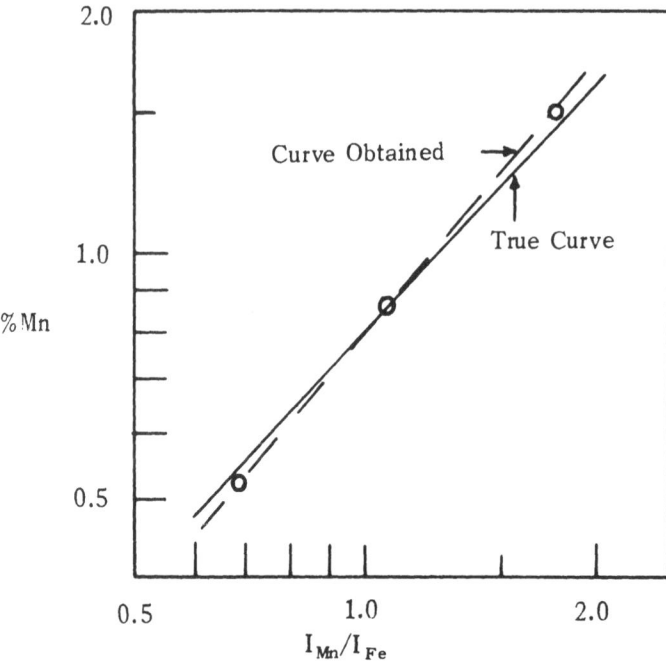

Fig. 1. Systematic error caused by deviation from the true curve.

cause an improper curve to be obtained, the curve once drawn is
always wrong. The obvious solution to this problem is to average
enough analytical results on each individual standard, so that the
*statistical* error is reduced to an acceptable value and the curve obtained becomes, in reality, the true curve. The statistical error of the
average of repeated runs on a sample is inversely proportional to the
square root of the number of values. Thus, averaging four values re-
duces the error to one-half; nine, to one-third; and sixteen, to one-
fourth. You can see that improvement becomes increasingly expen-
sive and that the range from two to nine values is the most
remunerative. When one determination is to be used in analyzing unknown samples, a good compromise for determining the analytical
curve is to average four values on each of three standards.

To obtain a good approximation of the "true" analytical curve it
is apparent that the standards must be homogeneous and accurate. If,
for example, the chemical value of the high standard in Fig. 1 is in
error, then this point will be in error even if the intensity ratio is
exactly determined. The composition range of the standards must

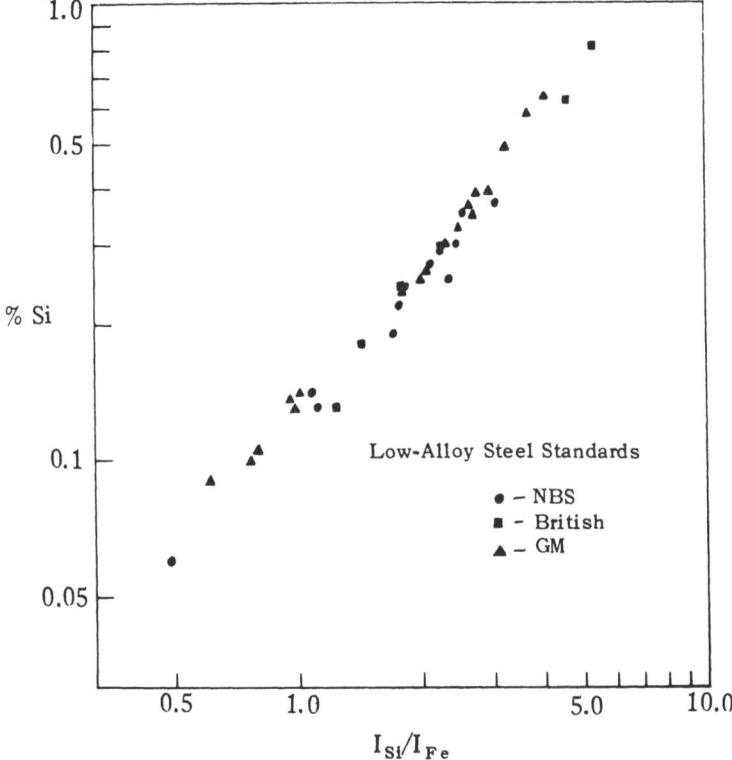

Fig. 2. Analytical results on 36 different low-alloy-steel standards.

also be adequate since the possibility of an error increases rapidly if a curve is extrapolated. At least three standards, two bracketing the desired range and one in the center, are required to determine an analytical curve adequately. The standard in the center is needed to determine the shape of the curve.

## DRAWING THE CURVE

Figure 2 illustrates two points: first, the variations that can be encountered even when primary standards are used; and second, the problem of drawing the best curve through a mass of scattered points.

Plotted here are the averages of four shots on each of 36 different low-alloy-steel standards. These consist of 11 standards issued by the National Bureau of Standards, 7 by the Bureau of Analyzed Samples,

Ltd., in Britain and 18 General Motors Spectrographic Standards. The analytical method used is known to be essentially free of matrix effects for low-alloy steel, and the width of each point is equivalent to the standard deviation for the average of four values. Thus, the scatter shown here is not primarily due to errors incurred in determining the Si-to-Fe intensity ratio, but rather to nonhomogeneity and chemical errors inherent in each standard. By using a great many standards, any bias or systematic error inherent in one particular set can be greatly reduced.

The collection of points shown presents an interesting challenge to the person responsible for drawing the curve. On the basis of similar curves, and because the data indicate no reasonable form for a nonlinear curve, it can be assumed that the curve lies in a straight line. Where would you draw the straight line?

An excellent solution to this problem is the method of least squares, which can be applied when statistical errors are nearly constant for all points. To make the statistical errors in emission spectroscopy constant, we plot the data on logarithmic coordinates and we let $x = \log$ intensity ratio and $y = \log \%$ Si. The least-square equations become:

$$\bar{x} = \Sigma x/n \qquad \bar{y} = \Sigma y/n$$

$$b_{slope} = \frac{\Sigma(y-\bar{y})^2}{\Sigma(x-\bar{x})(y-\bar{y})} = \frac{n\Sigma y^2 - (\Sigma y)^2}{n\Sigma xy - \Sigma x\Sigma y}$$

The least-squares line obtained with these equations has the property that the sum of the squares of the $x$ or log intensity ratio deviations from the line is smaller than for any other line. It also has the interesting and useful property that it passes through $\bar{x}$, $\bar{y}$, the grand-average point.

As you can see from the equations, the calculation of the slope can be a lengthy task, especially if there are a large number of points and no calculating machine is available.

Figure 3 contains the same points as Fig. 2 and a straight line drawn by an abbreviated least-squares technique. The fact that the least-squares line passes through $\bar{x}$, $\bar{y}$, indicated by the large cross, was used in drawing the line. A transparent straight edge was positioned on this point and rotated until a visual estimate of the best fit was obtained. Constraining the line to go through $\bar{x}$, $\bar{y}$ greatly

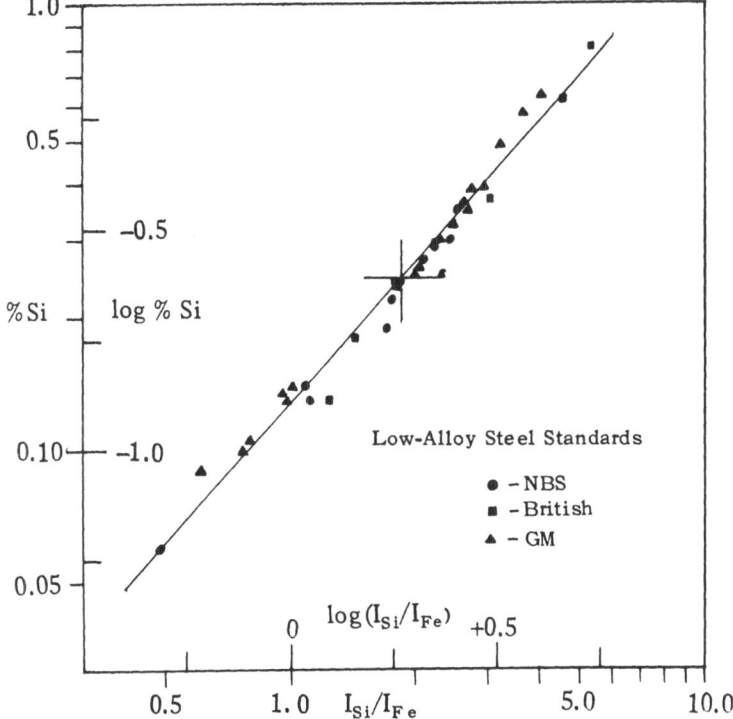

Fig. 3. Curve drawn using an abbreviated least-square technique.

simplifies and sharpens the visual estimate of the slope. With this technique only the $\bar{x}$ and $\bar{y}$ values need be computed.

The curve shown represents a best estimate for the determination of Si in low-alloy steel. If it became necessary to reestablish the curve at some later date because of changes in experimental variables, it would not be necessary to run all 36 standards again, but only three, a high, a middle, and a low. The chemical value used in replotting the curve would not be the certified value, but a value obtained by analyzing the three specific standards from the curve shown here.

## CONTROLLING CURVE SHIFT*

Day-to-day variations in analytical curves are a problem in emission spectroscopy, as indicated by a recent cooperative test in-

*The techniques for controlling curve shift were obtained from J. E. Jackson R. A. Freund, and W. G. Howe, "Errors Associated with Process Adjustments," *Virginia Journal of Science* 10, 1, 3–26, 2959.

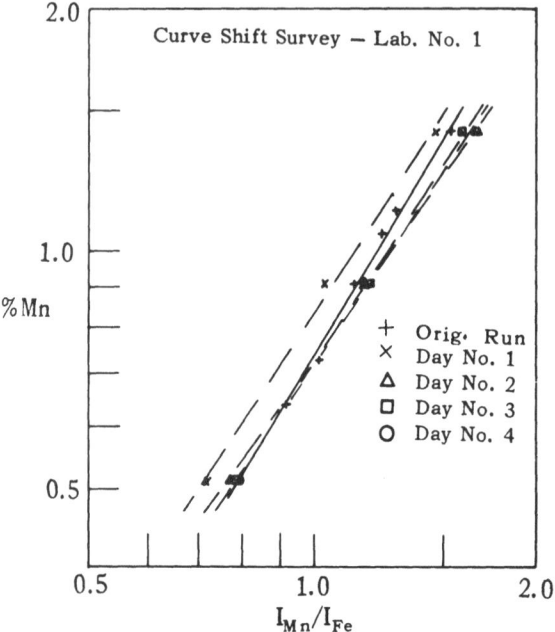

Fig. 4. An example of day-to-day curve shift.

volving 15 laboratories. Figure 4 shows some of the data from the test. In the cooperative test the original run consisted of obtaining three values on each of seven low-alloy-steel standards. On four subsequent days three of the standards, a high, a middle, and a low, were run four times. The size of the symbols represents the standard deviation for the points. The solid line indicates the curve obtained on the initial run of the seven standards and the dashed lines represent the magnitude of the day-to-day curve shift obtained with the high, middle, and low standards. As you can see, curve shift is indeed a problem. The first line of defense against curve shift is not statistics, but rigorous control over the experimental variables. Figure 5 shows a control chart obtained by running a standard sample on each plate. The minor maintenance job of cleaning the optics resulted in an appreciable curve shift which, if it had gone undetected, would have resulted in systematic analytical errors.

The technique used by some spectrographers of correcting all analytical results on a plate for the deviation indicated by a single standard sample on that plate is highly unreliable. It is tantamount to drawing a new curve based on one value. The random variations

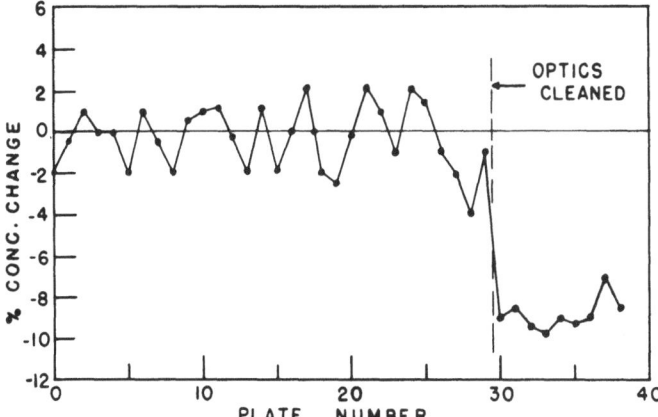

Fig. 5. Control chart showing effect of cleaning optics.

indicated by the spikes up to plate 29 in Fig. 5 would have caused plate-to-plate systematic errors if all the values on each plate had been adjusted by using the standard.

My personal feeling is that no run-to-run correction should be applied until there is reasonable evidence that a shift has occurred. When such evidence appears, then sufficient data should be obtained to reestablish the curve.

A procedure which fulfills these requirements follows:

1. Analyze a high and a low standard. Any curve for which both of the intensity-ratio values deviate from their expected values by less than two times the standard deviation shall be considered satisfactory.

2. For any curve where either or both of the values fall outside the two-standard-deviation limit, analyze the high and low standards an additional three times.

   a. If the means of the four readings on the two standards fall within one standard deviation of their expected values, consider the curve satisfactory.

   b. For any curve where one or both of the averages fall outside the one-standard-deviation limit, obtain sufficient additional data to draw a new curve.

This technique is especially applicable to direct-reading instrumentation, but cannot be conveniently used with photographic recording because the data cannot be obtained during the running of the plate. In the case of photographic recording, the following procedure might be used:

1. Analyze a high and a low standard on each plate. Consider satisfactory any curve for which both intensity-ratio values deviate from their expected values by less than one standard deviation.
2. For any point whose deviation from the expected value is from one to two standard deviations, use a correction of one-third the deviation and draw a new curve.
3. For any point whose deviation is two to three standard deviations, use a correction of two-thirds and draw a new curve.
4. For any point whose deviation is greater than three standard deviations, use a full correction in drawing the curve.

However, merely redrawing the analytical curve is not the ultimate solution. The experimental cause of the curve shift should be located and brought under control. The use of statistical methods of establishing and controlling curve shifts is no substitute for well-adjusted and maintained analytical equipment. In conjunction with experimental information, statistical techniques merely help in locating and controlling the important variables.

The analytical curve determines the accuracy of analytical methods and it behooves the analyst to be especially critical in establishing and controlling such curves.

# Purity of "National" Spectroscopic Electrodes

## J. Weinard

National Carbon Company
Division of Union Carbide Corporation

---

About 600 different spectroscopic samples (rods and preforms in L-113 SP carbon, AGKSP and SPK grades) were studied by spectroscopic quantitative analysis. The concentrations of impurities in parts per million were statistically compared with the visually estimated densities of the spectral lines obtained by the cathode-layer method. Concentration limits expressed as maximum concentrations in parts per million could be established for density estimations in the arbitrary scale from 0 to 3. Using the line-to-background ratio measured with the densitometer, a method was found of reporting impurities with their maximum concentration in parts per million with a confidence level of 95%.

## DISCUSSION

All manufacturers of carbon and graphite spectroscopic electrodes use the highly sensitive cathode-layer method as the standard analytical procedure for checking the purity of spectroscopic products. This method is qualitative, but gives an indication of the amount of the impurity present (high, low, absent) by estimating the density of the spectral line of an element in an arbitrary scale (x, not present; 0, barely visible; 1 to 6, visual density steps). This arbitrary scale is based solely on a subjective observation and estimation of the line density as seen on the spectrographic plates. The estimated density values are not representative of the concentration of a specific element. Different elements have different excitation energies. The densities of their spectral lines are, therefore, different for equal concentrations. Spectrographs, optical and electrical conditions, and type of emulsion are important factors which influence the estimations of the line density. Therefore, it is understandable that purity statements issued by different manufacturers may disagree considerably for spectroscopic products with exactly the same impurity levels.

The object of our investigation was the correlation of the density of the analytical spectral line to the real concentration value of a given element by the dc-arc external-standard method. The spectrograph used for the cathode-layer method is a Littrow Large Quartz Glass instrument built by Hilger. The conditions for the cathode-layer method are the following:

Arc gap . . . . . . . . . . . . . . . . . Approximately 10 mm
Arc current . . . . . . . . . . . . . . . Variable with size of electrodes
                                          (10–24 amp)
Exposure time . . . . . . . . . . . . . 30 sec, no preburn
Slit width . . . . . . . . . . . . . . . 20 $\mu$
Slit height . . . . . . . . . . . . . . . 10 mm
Plates . . . . . . . . . . . . . . . . . Kodak K33 in Q6 and Q3
                                          position
                                          Kodak IVN in the glass position
Developer . . . . . . . . . . . . . . . D19, 4 min

The analytical lines used in this method for detecting impurities are the following very sensitive lines:

| Al | 3082.155 A | Mn | 2596.104 A |
|----|------------|-----|------------|
|    | 3092.713 A |     | 2593.729 A |
| B  | 2496.778 A | K   | 7664.9 A |
|    | 2497.733 A |     | 7699.0 A |
| Ca | 4226.728 A | Si  | 2881.578 A |
| Cu | 3247.54 A  | Ag  | 3280.683 A |
|    | 3273.96 A  | Na  | 5890.0 A |
| Fe | 2483.27 A  |     | 5895.0 A |
|    | 2488.15 A  | Sn  | 2840.0 A |
| Pb | 2833.069 A | Ti  | 3239.0 A |
| Mg | 2795.53 A  |     | 3242.0 A |
|    | 2802.695 A | V   | 2908.8 A |

About 600 different samples were collected over a period of several months and analyzed by the cathode-layer method. The majority of these samples were $1/4$-in. spectroscopic electrodes. For the quantitative analysis of trace impurities, these samples were carefully powdered and analyzed spectroscopically by the external-standard method. The external-standard method was chosen to eliminate possible contamination by mixing an internal standard with the sample. Copper was used as the external standard. For the copper determination, cobalt served as the external standard.

The spectroscopic conditions for the quantitative analyses are listed below:

Spectrograph . . . . . . . . . . . . . . . . . Jaco Ebert, 3.4-m grating
                                              15,000 lines/in.
Slit . . . . . . . . . . . . . . . . . . . . . . . $20\mu$
dc arc . . . . . . . . . . . . . . . . . . . . . . 5 amp
Arc gap . . . . . . . . . . . . . . . . . . . . . 3 mm
Exposure . . . . . . . . . . . . . . . . . . . . 15 sec—2 sec preburn
Plate . . . . . . . . . . . . . . . . . . . . . . Kodak 103-0 (2200–4400 A)
                                              Kodak 1 N (4400–8000 A)
Developer . . . . . . . . . . . . . . . . . . . D19, 4 min
Electrodes . . . . . . . . . . . . . . . . . . . Highest purity, National
                                              L-4230, L-4242, and L-3754

Spectroscopically pure standards from Johnson Matthew, London, England, were used to establish the working curves for 15 elements. By diluting these standards with equal parts of highest-purity graphite powder (SP-2) in successive steps, samples were prepared with concentrations of 1, 5, 10, and 50 ppm for the less sensitive elements, and 0.1, 0.5, 1.0, 5, and 10 ppm for the sensitive elements (0.01 ppm for Mg) in order to construct the working curves. These synthetic standards in higher concentrations (0.1% and 2%) were analyzed chemically and the results were in excellent agreement with the calculated values.

In the quantitative analysis, the same analytical lines were used as reported for the cathode-layer method. For each sample, three spectra were taken and the average concentration for each element determined. These values were then compared with the line-intensity estimations. The spread of the concentration values was high for a given estimated density in the arbitrary scale. For estimations above 3 in the arbitrary scale, this spread was up to ±1000% and, therefore, was not investigated further. In density estimations of 0 to 3, the maximum spread of concentrations was found to be ±200%. Taking this into account, the density estimations of the analytical lines of the most common elements found as impurities in the spectroscopic products were correlated to the concentrations. An example for two elements, aluminum and silicon, is given in Table I.

**TABLE I**
Analyses of Aluminum and Silicon

| Element | Number of samples | Estimated density of analytical line | Measured maximum concentration, ppm | Number of determinations |
|---|---|---|---|---|
| Aluminum | 110 | 0 | 0.5 | 330 |
| | 105 | 1 | 2.0 | 315 |
| | 68 | 2 | 10.0 | 204 |
| | – | 3 | Too-high spread | – |
| Silicon | 111 | 0 | 1.0 | 333 |
| | 96 | 1 | 3.0 | 288 |
| | 63 | 2 | 10.0 | 189 |
| | – | 3 | Too-high spread | – |

As seen in Table I, 110 samples were investigated for aluminum which was analyzed by the cathode-layer method with a density estimation of zero. Each sample was arced three times for a total of 330 determinations. The spread found in the 330 determinations was as follows: the spectra didn't show aluminum in 17 (5%) of the samples; the spectra gave an aluminum concentration of between 0.1 and 0.2 ppm for 56 (17%) determinations; between 0.2 and 0.3 ppm for 89 (27%) determinations; between 0.3 and 0.4 ppm for 109 (33%) determinations; between 0.4 and 0.5 ppm for 49 (15%) determinations; and between 0.5 and 0.6 ppm for 10 (3%) determinations.

Figure 1 gives the distribution of the concentration values of aluminum for cathode-layer-density readings of zero. With a confidence level of 95%, the maximum concentration of aluminum in these samples was found to be 0.5 ppm.

Figure 2 shows the distribution of the concentration values of silicon for cathode-layer-density readings of 1. In this case, the maximum concentration with a confidence level of 95% would be 3 ppm Si.

These investigations were made for 15 elements listed in the purity statements: aluminum, boron, calcium, copper, iron, lead, magnesium, manganese, potassium, silicon, silver, sodium, tin, titanium, and vanadium. In order to evaluate maximum concentrations of all these elements it was necessary to take samples rejected from purified lots because of high impurities. Normally, only three to five elements are detected after a satisfactory purification. The distribution curves of the concentration values of all investigated elements were similar to the curves of Figs. 1 and 2.

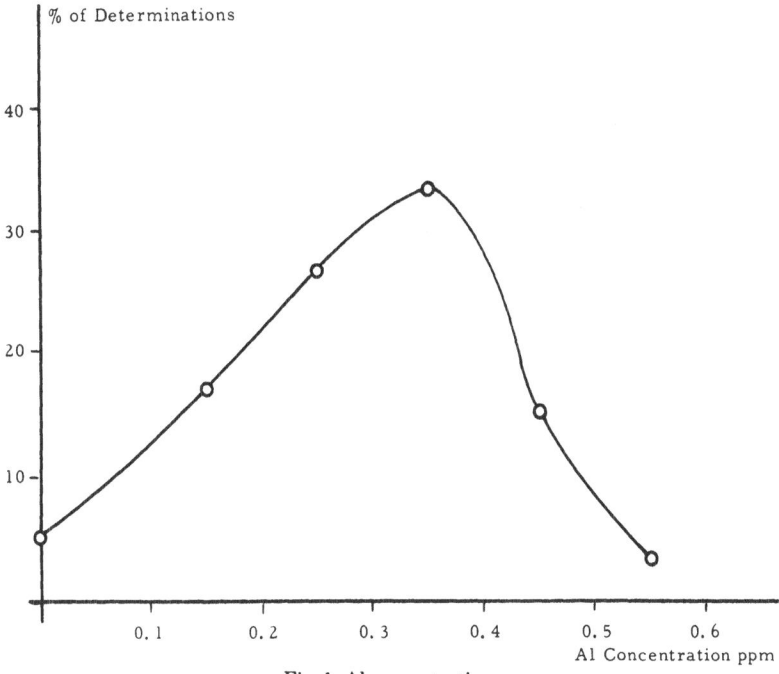

Fig. 1. Al concentration.

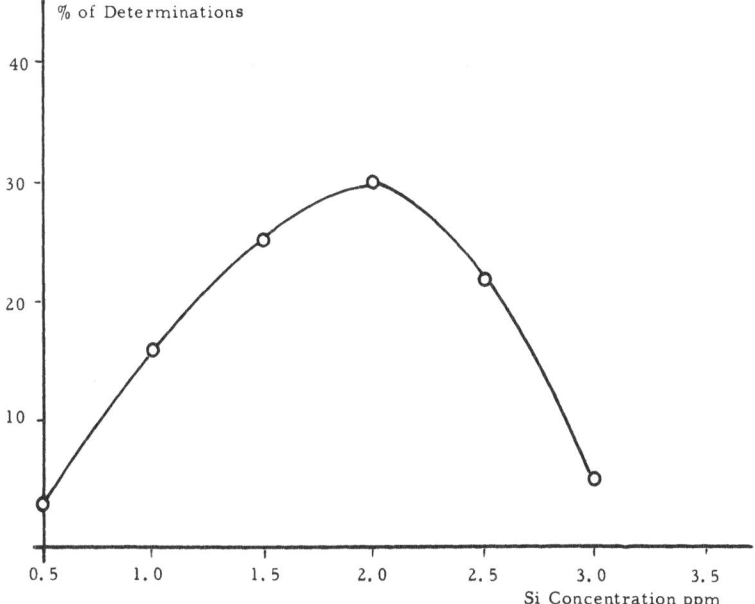

Fig. 2. Si concentration.

Visual line-density estimations are neither very accurate nor very reproducible, because of the human element. The density steps in the arbitrary scale cover a too-wide density range and a visual differentiation is impossible. However, with a great number of data at our disposition—about 9000 different spectra were evaluated—we were able to correlate the measured line-to-background ratio to the actual concentration of a given element, thus eliminating visual estimations. Again, this correlation was set up with a 95% confidence level. With this method, the purity of spectroscopic electrodes can be controlled to a much finer degree. Knowing the maximum concentration of impurities in the electrodes, the spectroscopist can approach the most critical analytical work with confidence that his results will not be affected by elements present in the electrodes in indefinite concentrations.

# Spectral Excitation with Stabilized Plasma Jets*

## Louis E. Owen

Goodyear Atomic Corporation
Portsmouth, Ohio

It has been agreed for some years that further improvement in the precision of spectrographic analysis must await improvements in the excitation process. Photoelectric spectrometers are so perfected that their error contribution has been considered negligible.

Stabilized plasma jets may now offer a new dimension in excitation precision equaling that of present-day spectrometers. These plasma jets, used for gas sample or solution excitation, exhibit an unusual steadiness for what is basically a dc arc.

Two models of the stablizied plasma jets are described. With one model, useful for isotopic assay systems, uranium spectra can be directly excited from gaseous uranium hexafluoride. The more generally useful model is adapted to the spectral excitation of solutions. In each case the sample is introduced into the center of a plasma discharge; the gas is introduced directly, the solution after atomization.

Stabilized plasma jets are characterized by high-temperature excitation, great brightness, large power-input capability, and steady-state operation.

## INTRODUCTION

For many years it has been apparent that further improvement in spectrochemical analysis must be based on improvements in the excitation process. Photoelectric spectrometers are perfected to the point where their analytical error contribution is considered negligible. A new approach to the excitation problem came with the proposal of Margoshes and Scribner [1] for plasma-jet excitation in spectrochemistry. In a plasma jet an arc started between two electrodes is blown out through an orifice in one electrode in the form

*This work was performed under contract AT-(33–2)-1 with the U. S. Atomic Energy Commission.

of a "flame." The same is introduced into the arc chamber and its emitted radiation is viewed in the flame discharge.

The defect of the original design—positional instability of the flame—was eventually overcome by the addition of an external electrode which serves as an electrical return [2]. The stabilized version of the plasma jet is especially suited to the direct excitation of uranium hexafluoride [3].

## GAS SAMPLE JET

The gas sample jet (shown in Figs. 1 and 2 and described in Table I) has two principal components which are insulated from one another by the plastic sleeve $D$. The main body of the jet is the water-cooled cathode assembly $C$ which encloses the discharge chamber and carries the graphite exit-orifice electrode $B$. Electrically, this cathode assembly is an integral part of the support rod and the external tungsten electrode $A$. It is at ground potential for dc-arc power sources which have a ground negative lead.

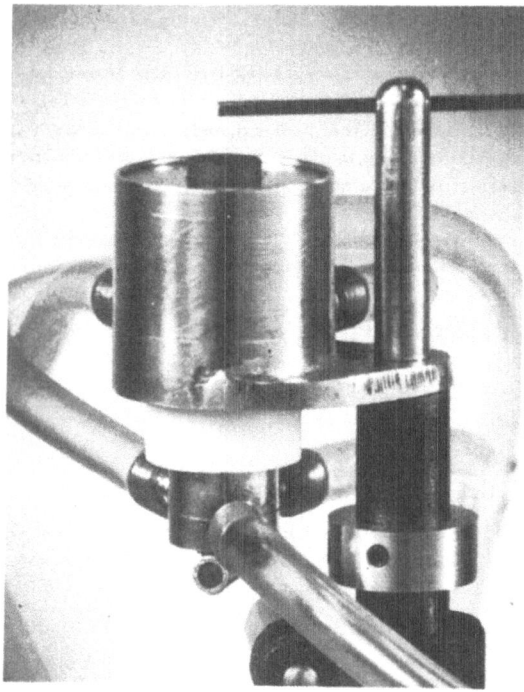

Fig. 1. Gas sample jet.

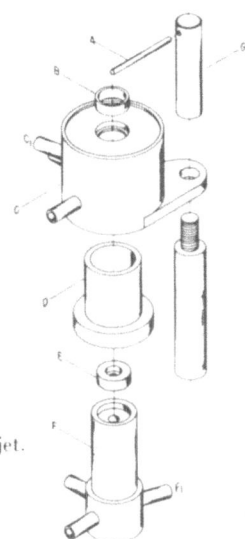

Fig. 2. Gas sample jet.

## TABLE I
### Components of the Plasma Jet for Gas Samples

| Key* | Description | Material | Dimensions |
|------|-------------|----------|------------|
| A | External electrode | Tungsten | 1/8-in. o.d. rod |
| B | Cathode electrode | Graphite | 12.5 mm o.d., 5 mm thick, 10-mm orifice |
| C | Cathode assembly | Brass | 1-1/2 in. o.d., 3/4 in. i.d. |
| $C_1$ | Tangential gas inlet | Copper | 3/16 in. o.d. |
| D | Insulating sleeve | Teflon | 3/4 in. o.d. 5/8 in. i.d. |
| E | Anode electrode | Graphite | 12.5 mm o.d., 5 mm thick, 5-mm orifice |
| F | Anode assembly | Brass | 5/8 in. o.d. 13/64 in. i.d. |
| $F_1$ | Sample gas inlet | Copper | 3/16 in. o.d. |
| G | External electrode holder | Brass | 3/8 in. o.d. |

*Key for Fig. 2.

The inner anode assembly $F$, which is also water cooled, carries the replaceable graphite anode electrode $E$ through which the sample gas is fed into the discharge.

## Operation

The arc initially forms between the two graphite electrodes but is quickly blown through the upper cathode electrode orifice in the form of a "flame." As the "flame" reaches the tungsten rod, the electrical discharge path transfers to the tungsten, and the graphite cathode electrode $B$ is effectively removed from the circuit. The orifice, therefore, is not subject to electrical erosion. The useful radiating column viewed by a spectrometer is the area 8 mm high and 3 to 4 mm in diameter between the tungsten electrode and the orifice electrode $B$.

The voltage drop across the jet is about 95 v for 12- to 28-amp currents when helium is introduced tangentially, at $C_1$, as the swirling gas of the jet. About 9 liters/min of helium are used during an excitation.

The $UF_6$ sample is frozen out on the walls of a trap (Fig. 3)

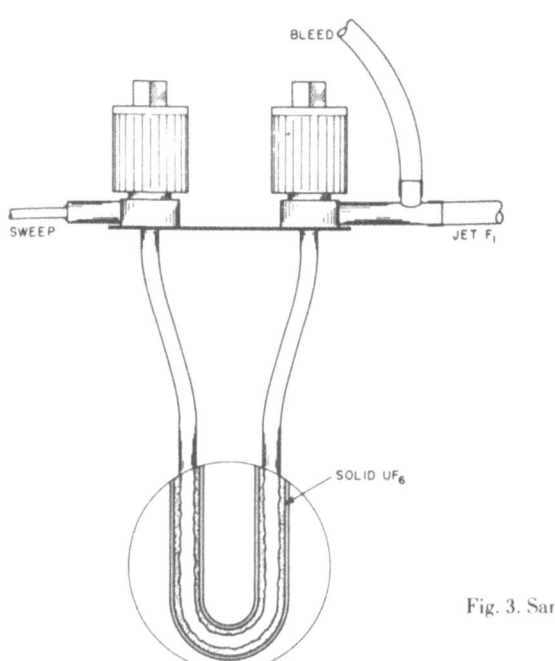

Fig. 3. Sample system.

by passing a gas stream containing $UF_6$ through a trap immersed in cold water. During excitation a modest flow of helium (2 to 10 ml/min) acts as a sweep gas, taking advantage of the vapor pressure of $UF_6$ at room temperature (about 100 mm Hg) to carry the sample into the jet. An additional flow of helium (labeled "bleed" gas in Fig. 3), at approximately 2 liters/min, enters the sample stream beyond the trap to dilute the $UF_6$ and to help carry it to the excitation zone. From 2 to 15 mg/min of $UF_6$ are consumed in average excitations. Tests in which the tubing and electrodes were not changed between samples of various assay levels indicate that the device does not exhibit a "memory" effect.

The external tungsten electrode is very gradually consumed and must be manually repositioned after an extended period of operation. Its position is not critical while the jet stream impinges upon it. If the tungsten rod burns away from the stream, an arc streamer forms from the "flame" to the rod and the voltage drop across the device increases markedly.

### Control

The sample trap mounts on the gas manifold (Figs. 4 and 5). At the start of an excitation, solenoid 1 is energized and helium flows into the manifold and onto the jet through the tangential and bleed lines. The sweep gas is initially blocked by the sample-trap inlet valve. The safety-pressure switch immediately closes and, through other controls, turns on dc power and an rf ignition pulse. The ignition pulse is automatically terminated as the arc current is established. The toggle valve which shunts the tangential needle valve is actuated during starting to hasten transfer of the cathode spot to the tungsten electrode.

The $UF_6$ sample is sent to the jet when the sample-trap valves are manually opened. When an excitation is to be terminated, the trap valves are closed and solenoid 1 is deenergized.

The bleed and sweep flow rates are set by the pressure drop across sections of capillary tubing and are not variable in routine operation. They determine the rate at which the $UF_6$ is introduced into the jet discharge. The tangential helium, however, is not completely preset but is manually varied over a small flow range during excitation. It is controlled by a needle valve which is adjusted to maintain discharge stability as monitored by an oscilloscope. The required variations in tangential flow rate do not appreciably affect the uranium emission.

Fig. 4. Sample trap.

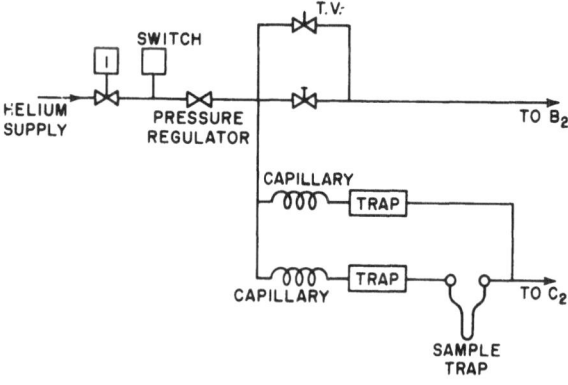

Fig. 5. Gas jet flow plan.

**Uranium Emission**

The optical structure of the jet flame, viewed as a source of uranium radiation, is that of an emitting cylinder surrounding a less-intense emitting rod. The least critical way to view the discharge is, therefore, also the simplist— in the center of the column. Results of vertical and horizontal measurement of signal-to-background (S/B) ratios also confirm this positioning.

The signal-to-background ratio for the $U^{235}$ line at 4244 A is not as favorable as from a hollow-cathode discharge, but is superior to that available in regular dc or ac-arc excitations.

The uranium spectrum is emitted so steadily by the jet that spectral lines can be recorded directly by scanning the spectrum over the exit slit of a photoelectric spectrometer. The spectrum in the 4244 A region is quite similar to that from hollow-cathode discharge lamps.

Optimum emission requires that a lower limit of $UF_6$ flow rate must be exceeded, because as the $UF_6$ in the discharge is reduced by reducing the sweep rate, the signal-to-background ratio of the $U^{235}$ line slowly improves until a limiting point is reached. After this point the background increases rapidly relative to the uranium line, and the ratio deteriorates. At greater flow rates the uranium suppresses the background radiation of the discharge and an optimum operating point can be established according to the intensity requirements of the spectrometer being used.

The intensity of the uranium spectra, independent of the sample flow rate, varies with the diameter of the orifice electrode (Fig. 6). The curve drawn is representative of a family of curves, each valid for a specific set of gas-flow parameters and jet geometry. The largest orifice practical with the jet in use is about 1 cm. The greatest intensity and the best signal-to-background ratio are obtained simultaneously with this orifice.

Increased sensitivity and improved ratios also occur when the tangential or bleed flow rate is reduced. In each of these instances, however, a limiting position of discharge instability occurs and it is necessary to back-off to a stable position. The unstable condition is dependent upon current level. This current dependency makes it impossible to prepare a simple graph showing uranium intensity versus current over the full current range of operation. The jet can be maintained with currents greater than 4 amp. At currents below 12 amp, however, the start-up arc transfer is frequently delayed.

Sensitivity and precision for the jet have been measured only
with the analytical spectrometer assembled at the Oak Ridge Gaseous
Diffusion Plant by Lee *et al.* [4]. On this instrument $U^{235}$ emission
in a sample of material known to contain only 0.038% $U^{235}$ was
unequivocally detected. The principal work with jet excitation is still
with material at the normal or 0.7115% level, since this is ubiquitous
and interlab comparisons subsequently are easy to make.

Precision tests using normal uranium made under conditions
comparable to plant stream analysis provided coefficients of variation
of 1% and less. While such coefficients of variation are not unusual
with photoelectric spectrometers, there are two special considerations: (1) this is as good as or better than hollow-cathode excitation
at this level, and (2) this precision was obtained with a signal-to-
background ratio of 1.1 to 1. A spectrum line with such a low S/B
would be considered below the limit of detectability in usual spectrochemical techniques. The precision obtained at such a disadvantageous ratio suggests that background with jet excitation is extremely
reproducible and is uranium-related rather than random noise.

The gas-fed plasma jet is a steady-state device. The reproducibility it provides is related directly to the effort spent in controlling
its sample and gas flows and the constancy of its electrical power.

### Status of Gas Jet

The plasma jet for exciting $UF_6$ is developing well as a component. Its special feature is the high radiation intensity obtainable

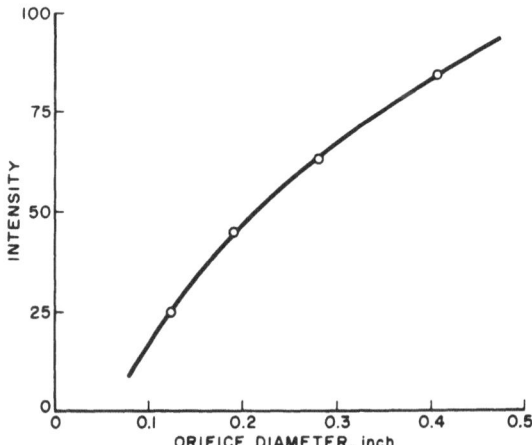

Fig. 6. Effect of orifice variation.

with low background. The jet should make it possible to assemble a direct-reading isotopic assay spectrometer which can be used as a plant on-stream instrument. The excitation subsystem, based on the jet, is being tested with various dispersive instruments and read-out circuits to evaluate the possible performance of complete analytical systems.

## SOLUTION SAMPLE JET

The stabilized plasma jet for solution samples is likely to be of more general applicability. Placing samples in solution not only obviates any original heterogeneity but also makes possible their non-selective introduction to an exciting discharge. Both of these points can be important, especially with highly alloyed metal samples. It was for such samples that Margoshes and Scribner first proposed plasma jet excitation [1].

### Construction

The stabilized plasma jet for solutions, while superficially resembling the gas jet model, differs in construction (Fig. 7). The principal components (Table II) are: $C$ the cathode assembly; $D$ the arc chamber enclosure; $F$ the anode assembly; $G$ the atomizer insulator; $H$ the atomizer; and $I$ the external electrode holder.

### Operation

The arc initially forming between the electrodes $B$ and $E$ is blown out through $B$ by helium which enters the arc chamber tangentially through $D_1$. The plasma stream impinges upon the external electrode A and the electrical return for the arc transfers to it. The sample is sprayed into the discharge column by the atomizer assembly $H$. Argon is used as the lift gas for the atomizer.

### Controls

The helium and argon gas flows to the solution jet are controlled by a solenoid-operated manifold (Fig. 8). To start an excitation, solenoids 1 and 2 are energized simultaneously. The helium goes through solenoid 1, a needle valve, and a flowmeter to the tangential gas inlet $D_1$ of the jet. The argon goes through solenoid 2, a needle valve, a flowmeter and, initially, through a bypass needle valve around solenoid 3 to the lift gas inlet $H_1$ of the atomizer. This initial flow of argon, while insufficient to lift the sample into the discharge zone, is sufficient to protect the atomizer tip from the

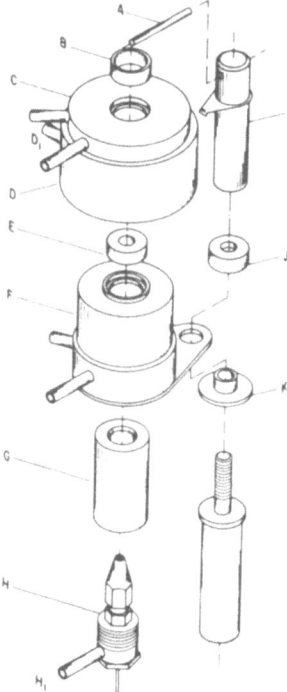

Fig. 7. Solution jet.

plasma. To get the sample into the discharge, solenoid 3 is energized. The argon flow increases so that it is capable of lifting the sample and spraying it into the discharge zone.

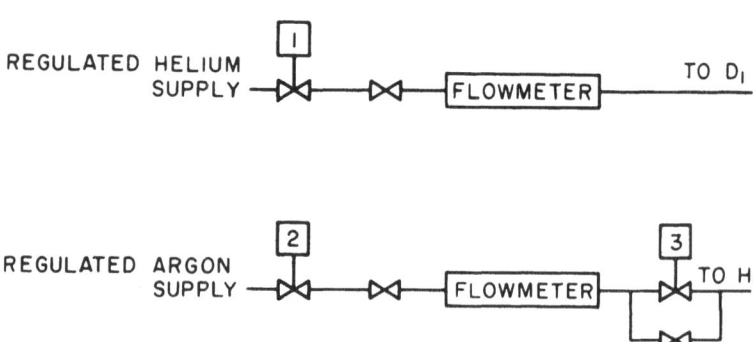

Fig. 8. Gas system for solution plasma jet.

## TABLE II
### Component Parts of Solution Plasma Jet

| Key* | Description | Material | Dimensions |
|------|-------------|----------|------------|
| A | External electrode | Tungsten | 3/32-in. o.d. rod |
| B | Cathode electrode | Graphite | 12.5 mm o.d., 10 mm i.d., 5mm thick |
| C | Cathode assembly | Brass | 1-3/4 in. o.d., 1/4 in. thick |
| D | Chamber body | Bakelite | 1-3/4 in. o.d., 1-5/32 in. i.d., 3/4 in. high |
| $D_1$ | Tangential gas inlet | Copper | 3/16 in. o.d. |
| E | Anode electrode | Graphite | 12.5 mm o.d., 5 mm i.d., 5 mm thick |
| F | Anode assembly | Brass | Outside diameter to fit inside diameter of chamber body, step edge to give desired separation of graphite electrodes, 5/8 in. i.d. |
| G | Atomizer assembly insulator | Teflon | 5/8 in. o.d., 3/8-in. −24 internal thread |
| H | Atomizer assembly | Brass | Inner part of Beckman Atomizer Burner |
| I | External electrode holder | Brass | 3/8-in. o.d. rod, inside threaded 1/4 in. −20 |
| J | Insulating washer | Bakelite | 5/8 in. o.d., 1/4 in. thick |
| K | Insulating washer | Teflon | 1 in. o.d., 1/4 in. thick |

*Key for Fig. 8.

The use of argon in the solution plasma jet results in a much lower impedance than in the gas jet. Consequently the solution jet is easier to power than the gas jet, which uses only helium.

## Sensitivity and Precision

While the subject has not been intensively investigated experience indicates that the sensitivity of the jet is comparable to other solution methods. Its special value is to be found in its stability.

The original National Bureau of Standards (NBS) plasma jet provided coefficients of variation of 1% to 1.5%, with most of the residual error being attributed to the erratic wandering of the anode and cathode spot. The development of the stabilized jet made it desirable to test this hypothesis, so the tests were rerun at NBS using the external electrode model. The results of these tests confirmed the value of stabilization as substantially lower coefficients of variation were obtained without optimization of operating parameters. The limiting factor appeared to be constancy of atomization. With control of this parameter, coefficients of variation of less than 0.5% should be readily attained.

## Status of Solution Jet

The stabilized plasma jet for solutions is an unusually precise high-temperature excitation device for spectrochemical use. It must be investigated and refined by those workers with specific need for its special attributes.

## REFERENCES

1. M. Margoshes and B. F. Scribner, *Spectrochim. Acta* 14, 2, 138, 1959.
2. L. E. Owen, "Stable Plasma Jet for Excitation of Solutions," *Appl. Spectroscopy* (in press).
3. L. E. Owen, "Plasma Jet Excitation of Uranium Hexafluoride," in *Proceedings of the Fourth Conference on Analytical Chemistry in Nuclear Reactor Technology. Gallinburg, Tennessee, October 12–14, 1960.* Washington, USAEC (TID-7606).
4. T. Lee, S. Katz, and S. McIntyre, *10th Conference on Applied Spectroscopy,* Pittsburgh, 1959, paper 143.

# Procedures for Testing Diffraction Gratings

## Edward Leibhardt

Diffraction Products, Inc.
Maywood, Illinois

The first part of the paper will cover, in outline form, methods for testing gratings from the standpoint of the buyer. These will include both the visual methods for testing, without mounting the grating, and also the photographic and photometric methods when the grating is suitably mounted.

The second half of the paper will cover tests that a manufacturer of gratings uses and how these are correlated to engine performance.

### ROWLAND GHOSTS

For this test a low-pressure mercury lamp such as a G. E. 4-w Germicidal Lamp can be used. The mercury spectrum is viewed in all available orders, using a narrow slit and an eyepiece. In modern gratings, the estimate of the relative intensity of the ghosts is best made in the higher orders, making use of the square law. This law states that the relative intensities of the ghosts are roughly proportional to the square of the order of the spectrum. One of the following methods can be used for estimating the relative intensity of the ghost.

1. If a camera is available in place of the eyepiece, a series of photographs having different lengths of exposure can be taken. A comparison of the density of the ghost as compared to that of the main line on one of the series of plates will give a fair estimate of the ghost intensity.

2. A rotating disc with a radial slot of adjustable width is used to reduce the intensity of the main line a known amount, until the intensity of the main line and that of the ghost are equal. This method is not very accurate. For example, with a slit width of 1 mm, the circumference of the disc would have to be 1 m in order to check

a grating whose ghost intensity was 1/1000 of the parent line. In most modern gratings the ghost intensity is much lower; for this reason this method could only be used in the higher orders.

3. A step filter with accurately known transmission could be used to cover up the main line and estimate when the ghost intensity matched that of the filtered main-line intensity. This is a quick method but, again, gives only an estimate.

4. The most accurate method is that using a photomultiplier tube and a strip-chart recorder with a suitable electronic system to give a linear response to intensity. Then either the grating is rotated or the slit is moved, and the peak response values of the main-line and its ghost, suitably amplified by a known amount, are recorded. For modern gratings it is sometimes necessary to amplify the ghost signal by as much as 10,000 times; for this purpose specially selected photomultiplier tubes with low background noise must be selected. Sometimes it is necessary to refrigerate the photomultiplier tube to get accurate readings. This test is usually made by the manufacturer of gratings, to keep a periodic check on the engine's performance and the accuracy of the end thrust bearing on the lead screw.

## LYMAN GHOSTS

Lyman Ghosts are characterized by the fact that they lie far from the parent line and originate from a periodic error not associated with the period of the screw; they can, therefore, easily be mistaken for a weak line. One of the easiest ways to check for these ghosts is to insert a filter which will pass only a very narrow band and examine the available orders for false lines. In modern gratings, Lyman ghosts are rarely found with an intensity that can be readily measured. A visual test for these ghosts can be made using a wide slit and a mercury source and observing the region between the zero-order and the first-order violet line. This will reveal Lyman ghosts if they are present. The mercury source must be intense or Lyman ghosts of less than 0.001 of the parent line will not be revealed. These ghosts are important and should be measured in the grating if it is to be used in the extreme ultraviolet region.

Another type of ghost which is sometimes called Lyman is the ghost caused by the index gears. For instance, in an engine where all the gear ratios are 1 : 1 for a standard 600-line/mm grating, no ghost will be found between the zero-order and first-order spectra. When the same engine rules a grating with 1200 lines/mm, it is necessary to use a 2 : 1 ratio of index gearing, and when the spectrum is examined, a weak 600-line first order spectrum will be present between zero order and the first order of the 1200-line/mm spectrum. The intensity of this spectrum can be examined visually using a clear tungsten bulb (200 w), and if any doubt arises, it can be checked photometrically. Usually, careful selection of the gears in the index train is all that is necessary to bring the value down to a safe value of 1/5000 of the parent line for a first-order spectrum. These ghosts are only troublesome if the grating is to be used in the ultraviolet region.

## RESOLUTION

The most frequent test for resolution is to photograph the hyperfine structure of the 5461 A mercury line, using a mercury arc lamp operating at a very low pressure. A photographic plate on which the line is exposed at low or medium density will reveal the line structure and from this the resolving power can be calculated.

Many other methods can be used for checking the resolving power, two of which are described below:

1. In the calcite crystal method devised by John Strong [1], one of the components of the mercury lines is split into two apparent components of opposite polarization by the calcite crystal, and the resolution is measured.
2. In a method devised by Rank, Shearer, and Bennett [2], the grating's ability to resolve the qui-intensity perpendicular components of a Zeeman triplet is observed.

## STRAY LIGHT

Disturbances in the diamond carriage or the diamond mounting give rise to minute errors which will influence the diamond as it rules a groove. These errors cause what is known as foot-light around strong spectral lines and this light is proportional to the square of the order. By far the greatest cause of stray light is scattering by the roughness of the groove edges of the grating at its surface. This is primarily caused by the diamond itself or the improper weight on the diamond causing an excessive amount of metal to be pushed up during the ruling of the grooves.

In replica gratings another and quite independent cause for stray light is found, namely, that caused by the plastic film. This film may, because of enclosure of dust particles, insufficient cure time, or a slow chemical reaction between the plastic and the aluminum layer, cause the replica grating surface to slowly deteriorate. This will cause an excessive amount of stray light to be scattered from the surface of the grating. It is for this reason that Harrison, Loof, and Loofbourow [3] found that original gratings have less stray light than replica gratings. This is also why manufacturers of replica gratings constantly strive to find more stable plastics.

Gratings to be used in the visual and near-ultraviolet region can be tested for common stray light in the following way. The grating is illuminated with a tungsten lamp (color temperature, 2800°K) at normal incidence. The ultraviolet radiation is absorbed by a filter so that no radiation from the second order overlaps the first order below 6000 A, the filter absorbing all radiation below 3000 A. A didymium filter (Corning 5120) is inserted in front of the source and the intensity at the 5825 A absorption band is measured with a photomultiplier tube.

If the spectral transmission of the didymium glass filter is small, the amount of stray light is obtained as the ratio of the measured currents, multiplied by a correction factor for the light which is incident through the filter on the grating, to the light that the unfiltered lamp would have given [4].

## TARGET PATTERN

If a photographic plate is inserted approximately halfway between the focal plane and the grating, a pattern of the light which would have formed the spectral line will be photographed. The regularity in the pattern, or the lack thereof, is a direct check on the performance of the ruling engine. The lack of regularity can generally be attributed to slight changes in the friction between the grating carriage and its ways, changes in the temperature of the ruling engine room, changes in the diamond ruling point, or a combination of all these plus more subtle changes which become apparent only after careful study. This test can also be done visually to some extent by placing the eye at the focal plane and looking directly at the illuminated area of the grating responsible for the spectral lines.

This visual method can be greatly increased in sensitiveness by the introduction of a knife edge (Foucault test) held in the focal plane. The image of the spectral line receives light from all areas of the grating if the grating spacing is uniform throughout. If there are inequalities in the spacings, the region of larger and smaller spacings send light slightly to one side or the other of the image produced by the average spacing. As the knife edge is introduced at the focal plane, areas with less than average spacings will darken first, depending on the direction in which the light is cut off. In the ideal grating, the light should vanish at the same time over the whole ruled area. Areas which stand out brightly will cause satellites to appear near the main line. This method can be directly correlated to the performance of the ruling engine, although it may not be easy to locate the cause for minute changes in the spacing without careful study of the ruling mechanism.

## INTERFEROMETRIC TESTS

There are two interferometric test methods in use for checking diffraction gratings by manufacturers of gratings. The first is the microinterferometer, which is used primarily for checking the quality of the individual ruled groove; in this way it gives a direct check on the diamond surface being used for the ruling operation and also a check of the angle of the blaze. The second is the interferometric test devised by Stroke [5], for showing the irregularities of the wavefront of a ruled grating, using a large-aperture Michelson interferometer modified in the Twyman-Green manner. The degree with which errors of groove spacings can be interpreted varies proportionately with the order or the sine of the angle at which the grating is illuminated; tests made in the sixth order are six times as sensitive as the phase-contrast method used by Ingelstam and Djurle [6], and it is much simpler to correlate with engine performance.

## REFERENCES

1. J. Strong, *J. Opt. Soc. Am.* 41, 1, 3–15, 1951.
2. Rank, Shearer, and Bennett, *J. Opt. Soc. Am.* 45, 762, 1955.
3. Harrison, Loof, and Loofbourow, *Practical Spectroscopy*, New York.
4. *Arkiv för Fysik* 3, 6, 63–81, 1951.
5. Stroke, *J. Opt. Soc. Am.* 45, 30, 1955.
6. E. Ingelstam and E. Djurle, *J. Opt. Soc. Am.* 43, 572, 1953.

# The Use of Time-Resolved Spectroscopy in the Investigation of Electrode Phenomena in Sparks

B. W. Joseph and R. F. Majkowski

Research Laboratories
General Motors Corporation
Warren, Michigan

The appearance of spectral lines in a spark in approximately $10^{-8}$ sec after the spark begins indicates the possibility that some other method besides electron heating is responsible for sample vaporization. Somerville postulates a process similar to sputtering as a possible mechanism for sample vaporization. Experimental evidence shows metallic ions are excited only when the sample is the cathode during an oscillating discharge. Although these time-resolved spectra do not as yet uniquely explain the vaporization mechanism, the evidence supports Somerville's postulate. Equipment for time-resolved spectroscopy is relatively uncomplicated and could be added to an existing spectrographic laboratory at reasonable cost. Additional research in this field will ultimately aid spectroscopy as an analytical tool.

## INTRODUCTION

The term "time-resolved spectroscopy" probably brings images of complicated electronic and/or optical equipment to the mind of the reader. This picture is not entirely true. Any spectrographer who has made a moving-plate study of an arc or spark has resolved the discharge in time. If you want to know the relative volatilization rates of materials in the sample, you look at, say, 5-sec sections of a 60-sec exposure. If you have enough intensity, you could crank the plate racker a little faster and look at 1-sec sections of the burn. Time-resolved spectroscopy in this part of the time spectrum has been used since the earliest days of spectrographic analysis.

Fig. 1. General view of optics for producing microsecond-resolution spectrograms. Three portions of typical spectra are shown at the right. The spectra shown are of a nickel alloy. The top one was obtained in air, the middle one in oxygen, and the lower one in argon.

The time-resolved spectroscopy to be described in this paper is a logical extension of moving-plate studies. If we can already look at a 1-sec section of the exposure, then by moving the plate quite a bit faster we could look at a single spark discharge lasting only 100 $\mu$sec or so. Rather than moving the plate, we can equivalently move the source along the length of the spectrograph entrance slit. For looking at a 100-$\mu$sec time section, rotating mirror optics such as those shown in Fig. 1 can be used. The rotating mirror is seen as the blurred object about in the middle of the picture. The type of spectra obtained is shown on the right side of the picture.

Though the equipment for looking at 100-$\mu$sec portions of the discharge is necessarily more complicated than moving plates, it can be stated to be simply an extension of moving-plate studies. In order to differentiate the two ends of the time spectrum, I would like to call studies of short-period phenomena "microsecond-resolution spectroscopy."

Since the early days of Bunsen and Kirchhoff, spectroscopy has constantly improved. The improvement in analytical technique has been directly related to the increase in knowledge of what goes on in the analytical gap. Moving-plate studies have been important in gaining an insight into some of the factors affecting the emission of light from the analytical gap. The studies of short-period phenomena which began with Schuster and Hemsalech [1] around 1900 have continued to interest the physicist and electrical engineer. Recently, Arpad Bardocz [2,3] has revitalized an interest in short-period phenomena as applied to spectrographic analysis, and has led us to study source parameters and atmospheres as they affect these phenomena in sparks.

## ELECTRONIC AND OPTICAL EQUIPMENT

In the present work, the technique used is only one of a number of ways of obtaining microsecond resolution of spectra. The equipment is essentially an evolution which is particularly suited to studying sources similar to those used in spectrographic analysis.

Since the time of Schuster and Hemsalech, a number of investigators have used Kerr cells [4-7] to resolve the spectrum in time, either visually or photographically. More recently, Blitzer and Cady [8] used a thyratron-triggered source and rotating mirror optics to obtain microsecond-resolution spectra of single condensed spark discharges. Photography of single discharges requires energies of the order of 100 joules. These energies are about ten ten times higher than those used for spectrographic analysis. By using photomultipliers gated synchronously with the discharge, Crosswhite, Steinhaus, and Dieke [9] were able to record the time spectra electronically, using a source of 1 to 10 joules, or in the range of normal spectrographic sources. The electronically triggered spark source developed by Bardocz is sufficiently repeatable in initiation time so that thousands of discharges may be accurately superimposed in time. Except for the triggering method, this source is identical to many contemporary sources in spectrographic laboratories and, for this reason, was used for the present work.

Figure 2 is a schematic of the optical system. This is a view looking straight down onto the system. As the mirror rotates, light from the triggering lamp strikes the phototube and fires the source. The light from the analytical gap passes through a horizontal slit which is imaged on the entrance slit of the spectrograph by a fixed concave

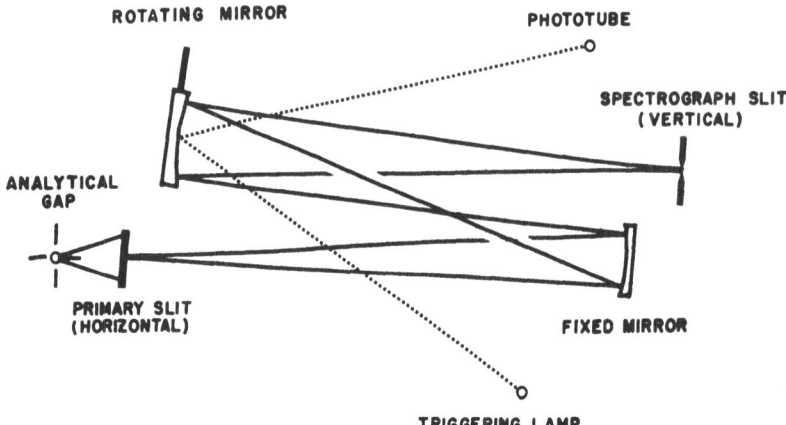

Fig. 2. Schematic rotating mirror optical system. The solid lines indicate the light path for the spark radiation, while the dotted lines indicate the light path for the optical trigger.

mirror and the rotating mirror in series. The fixed mirror folds the optical path and presents a diminished image of the horizontal slit to the rotating mirror. The height of the image of the horizontal slit on the slit of the spectrograph and the angular velocity of the rotating mirror determine the time resolution. In this work, the horizontal slit was used for time resolution of the spark and not primarily for spatial resolution of the gap. The spectrograph is a two-Cornu prism type with an aperture of *f*12 and a resolving power' of about 30,000 at 3000 A. For work in various atmospheres, the samples are enclosed in a simple 3-liter glass chamber with a quartz window.

Figure 3 shows a simplified schematic of the discharge portion of the source. With the capacitor $C$ charged to some voltage, the potential divider $R1$-$R2$ causes about half the voltage to appear across each of the auxiliary gaps $G1$ and $G2$. The gaps are set to be just on the verge of firing. As long as the thyratron is biased to cutoff, it appears as an open circuit across $G2$. When a signal to the grid of the thyratron causes it to conduct, it appears as a short circuit across $G2$. The full potential on the capacitor now appears across $G1$ and causes it to break down. Immediately that $G1$ breaks down, the potential across it drops to about 50 v. The full potential on the capacitor now appears across the analytical gap $G3$, breaking it down. After $G3$ breaks down, the discharge current flowing through the thyratron and its associated resistance causes the potential to

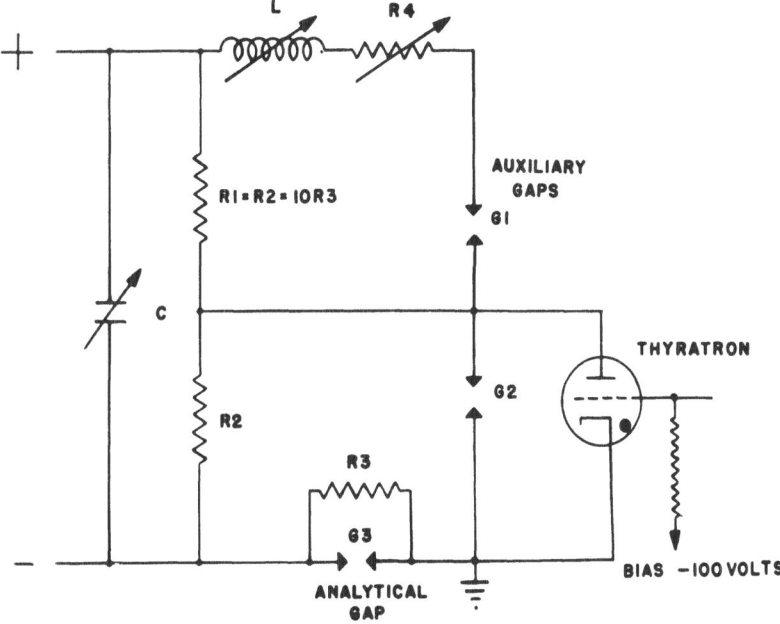

Fig. 3. Simplified schematic of the discharge portion of the triggered source.

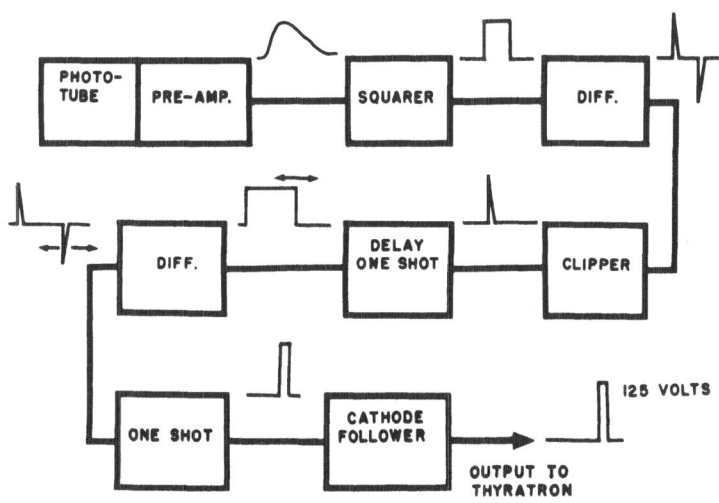

Fig. 4. Block diagram of optical pulse generator.

appear across $G2$, which then breaks down. The full discharge current then flows through all three gaps in series so that the thyratron does not carry the full load current. The source produces a typical oscillating discharge. The period of oscillation of the spark transient is determined by the inductance and capacitance of the discharge circuit, as in conventional sources.

The firing signal for the thyratron is derived from the rotating mirror through the pulse generator shown in Fig. 4. The signal from the phototube is shaped with a squaring circuit. The pulse is differentiated and clipped so the leading edge triggers a delay one-shot multivibrator. The output of the delay one-shot is a pulse of variable length, the leading edge of which is coincident in time with the leading edge of the light pulse. The variable-length pulse is then differentiated and the trailing edge used to trigger a second one-shot multivibrator. The output of the second one-shot is a positive pulse of 125 v at a given time after the light pulse. The cathode follower is used as an impedance matching device to the thyratron grid. The drift in the firing signal to the thyratron was verified to be less than 0.5 $\mu$sec in 1 hr, as referred to the light pulse. Short-term jitter is of the order of 0.06 $\mu$sec.

## EXPERIMENTAL RESULTS

Figure 5 shows two wavelength regions of a microsecond-resolution spectrogram of a nickel alloy. The left side is around 2800 A and the right side is around 2400 A. In this spectrogram and the ones to follow, wavelength appears horizontally across the figure and time appears vertically. The time resolution on these plates is 1 $\mu$sec. The period of oscillation of the spark transient in Fig. 5 is 4.7 $\mu$sec and the polarity of the discharge is such that the carbon counterelectrode starts out as the cathode. Lines from the atmosphere, air in this case, appear at the very beginning of the discharge. These persist only during the first half-cycle of the transient. These lines appear in the electron continuum. Next to appear are the lines of carbon from the counterelectrode. The first intensity maximum of these lines appears well within the continuum, and the intensity of these lines oscillates with a period equal to that of the spark transient. In the second half-cycle, the ion lines of the sample material appear and continue to oscillate with a period equal to that of the spark transient, but 180° out of phase with the carbon

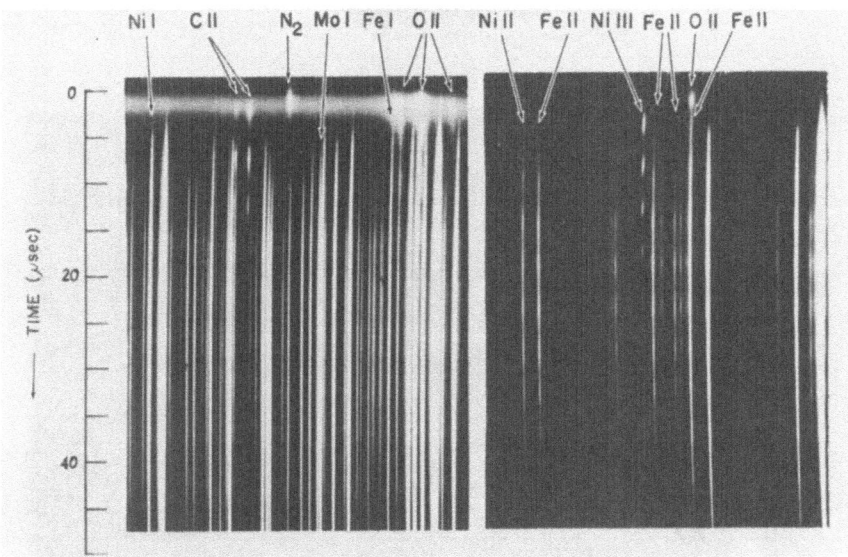

Fig. 5. Two wavelength regions of microsecond-resolution spectrum of a nickel alloy. Source parameters were: $C = 0.015$ farad, $L = 40$ h, $R = 0$, $I = 9.5$ rf amp. The exposure was 60 min in an air atmosphere. Resolution on this plate is 1 $\mu$sec.

lines. Typical arc lines of the sample, Ni I, Fe I, and Mo I, appear continuous in time with a single broad maximum.

The phase shift between the sample lines and the carbon lines is shown more clearly in Fig. 6. This is a portion of the copper spectrum. The period of oscillation of the spark transient is 9.3 $\mu$sec. Again, the polarity is such that the carbon counterelectrode starts out as the cathode. It is apparent that the copper line exactly fills in the spaces in the carbon lines. Since the carbon started out as the cathode, the carbon lines lead the copper lines. If the polarity is reversed, the copper lines lead the carbon lines. At the very beginning of the discharge there is some very faint radiation from the copper line.

A similar phenomenon is shown in Fig. 7. This is a portion of the aluminum spectrum around 3600 A. Again, the polarity is such that the carbon starts out as the cathode, and the period is 9.3 $\mu$sec. In comparing this figure with Figs. 5 and 6, it would appear that we have made a mistake. It certainly looks as if the aluminum is present right at the beginning of the discharge along with the nitrogen and $N_2$ from the air, and it looks as if the period of the discharge is

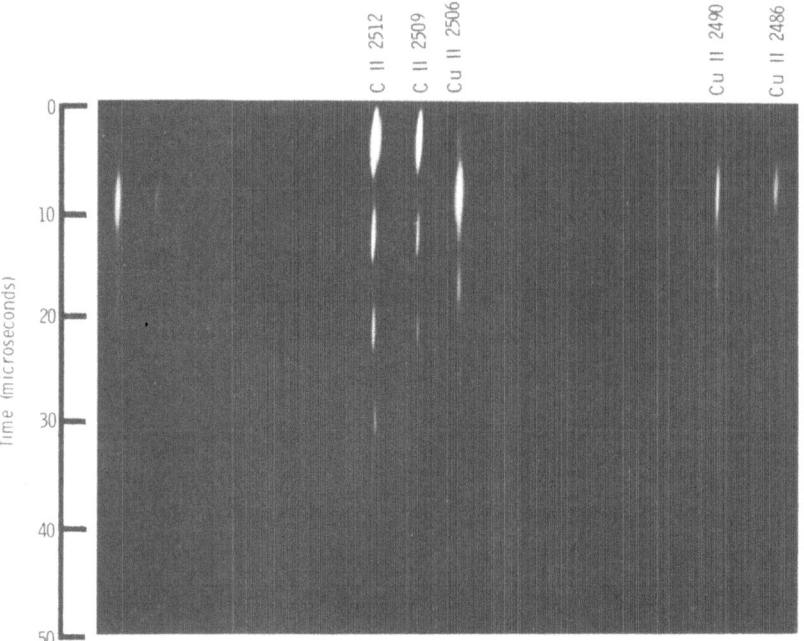

Fig. 6. Portion of microsecond-resolution spectrum of copper.

4.7 and not 9.3 μsec. Actually, there is no mistake. This aluminum was run with exactly the same conditions as the preceding copper. If you look closely at this aluminum spectrum, you will see that the second, fourth, sixth, eighth, and tenth half-cycles, during which the aluminum is the cathode, are indeed stronger than the odd half-cycles and the period of oscillation is truly 9.3 μsec. Although this Al 3586 A line is much too intense to make quantitative measurements, it is obvious that even though there is intense radiation when the sample is anode, the intensity is much higher when the sample is cathode. The appearance of the microsecond resolution spectrum of aluminum is significantly different than the spectra we have observed of nickel, copper, and iron. These differences that we observe point out one field for further investigation.

Information derived from plates such as shown in Figs. 5, 6, and 7 has helped to clarify the picture of the formation of the spark and electrode phenomena in sparks. Fassel and Tabeling [10] have shown that in unidirectional arcs, material is vaporized from the sample primarily when the sample is cathode. Loeb and Meek's

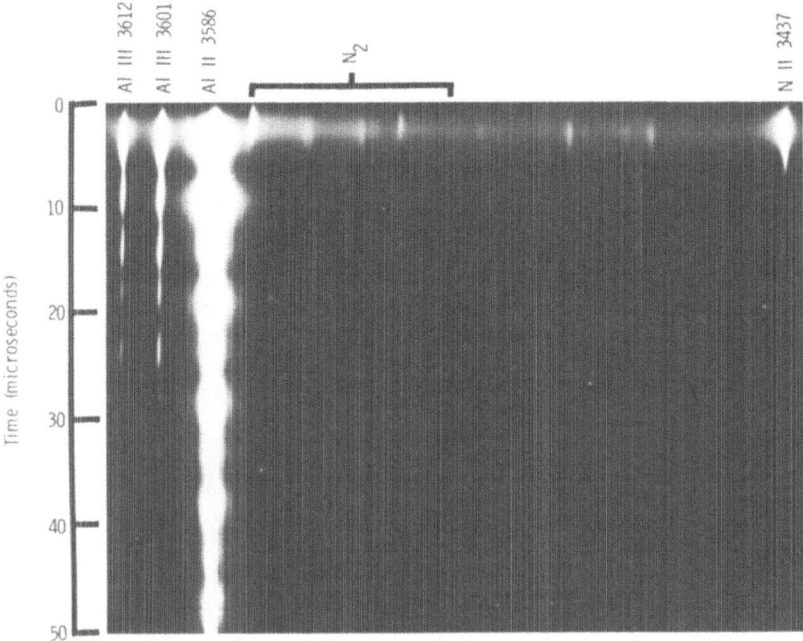

Fig. 7. Portion of microsecond-resolution spectrum of aluminum.

streamer theory of spark formation [11] postulates a stream of positive ions growing back to the cathode from the anode during the very first stages of spark formation. Studies of the spark channel growth indicate that in this type of discharge the current density is of the order of $10^6$ amp/cm$^2$. On the basis of this information, A. J. Rich [12] has calculated that resistance heating would only be expected to contribute 10 to 20% of the energy input to the sample. J. M. Somerville [13] postulates the direct action of positive ions on the sample surface to explain the appearance of spectral lines in much less time than it would take to boil the sample if resistance heating alone is responsible.

If we again look at the nickel spectrum in Fig. 5, we see from the high intensity of the O II lines that there are a very large number of ions in the atmosphere at the very beginning of the spark. Also, in the first half-cycle, during which the carbon is the cathode, the C II lines make their appearance very quickly. H. V. Knorr has measured the time of first appearance of electrode spectral lines to be something less than $10^{-8}$ sec after the spark begins. Although we do not have anywhere near this time resolution, we would estimate the time for carbon lines to appear would be of this order of magnitude.

Although our data are few and have not been quantitatively analyzed, they support the hypothesis that most of the sample vaporization is due to the direct action of positive ions in a manner similar to sputtering. We would concur with Blitzer and Cady that the streamer theory of spark formation is probably true. Even through we can support some of the theories on the formative stages of the spark, our understanding of the processes once the spark channel has formed are still rather hazy. The relation of process that we observe to those which are observable in spectrographic analysis are hazier still; yet, we are gaining a better understanding of what goes on in the gap through the use of microsecond-resolution spectroscopy.

For example, Fig. 8 shows the relative intensity of the Ni I 3064.4 A line as a function of time for various atmospheres. We already know that the atmosphere affects the line intensities as seen in normal spectrographic analysis [14]. Now we see that the intensities are affected in quite a profound manner when we look on the microsecond time scale. If we look at the effect of argon, for instance, not only is the peak intensity much higher than in air, but the line actually persists for a much longer time than in air. In fact, the lines persist for such a long time that we could not get the whole time range on the plate. Whereas in the other gases the lines have disappeared by 80 $\mu$sec or so, with an argon atmosphere they continue for 120 $\mu$sec or longer. Obviously, the integrated intensity observed in spectrographic analysis will be much higher in argon than in air. On the microsecond time scale we see that there are a number of factors which contribute to the higher intensity seen on the other end of the time scale. Now we have a better idea of the effect of argon on line intensities as seen in spectrographic analysis, and in addition we have some further clues to help determine the mechanisms responsible for this higher integrated intensity. By studying the effect of various gases on the short-period phenomena in the spark, we will eventually gain an insight into the mechanisms of sample excitation.

As another example, consider the ratio of a spark to an arc line. This "spark-to-arc ratio" has crept into the spectrographer's jargon and has been bandied about for some time. It has been found to be a sensitive indicator of sample excitation [15, 16]. It has been proposed as part of a method for specifying source parameters. But what does

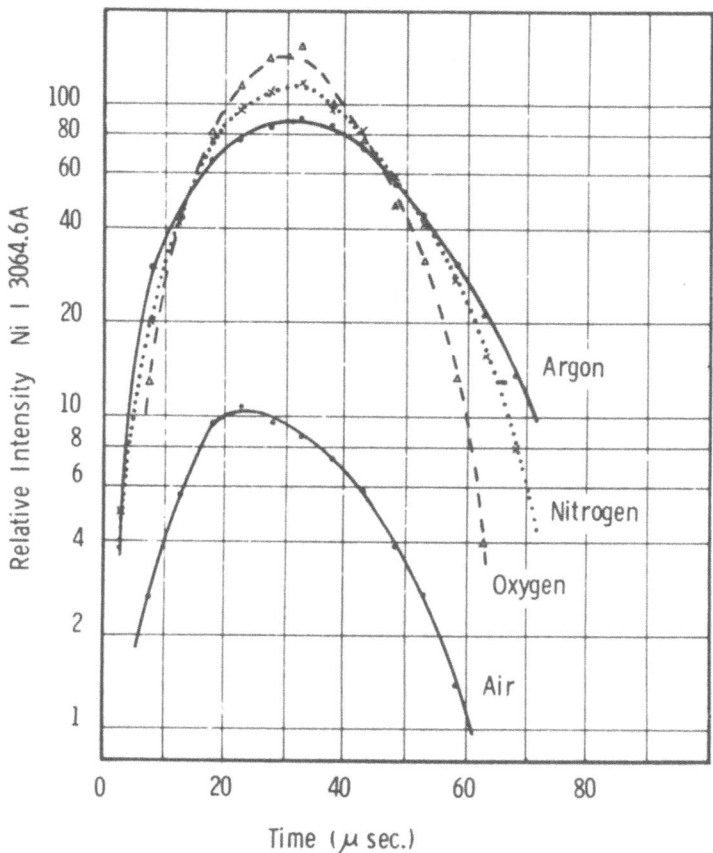

Fig. 8. Intensity variations of a nickel arc line during a spark for various atmospheres.

this ratio mean when the spark lines occur only on the current peaks when the sample is the cathode, while the arc lines appear much more continuous in time? We don't propose to answer that question, but we would like to show what we observe happening during the microseconds lifetime of a spark.

Figure 9 shows the Ni III 2448.4 A: Ni I 2472.1 A ratio as a function of time for various atmospheres. Since the Ni III line is discontinuous in time, we could only make microphotometric measurements on the peaks, and the abcissa of the graph is, therefore, in terms of cycles. The immediately obvious characteristic shown by these curves is the logarithmic decay of the III : I ratio.

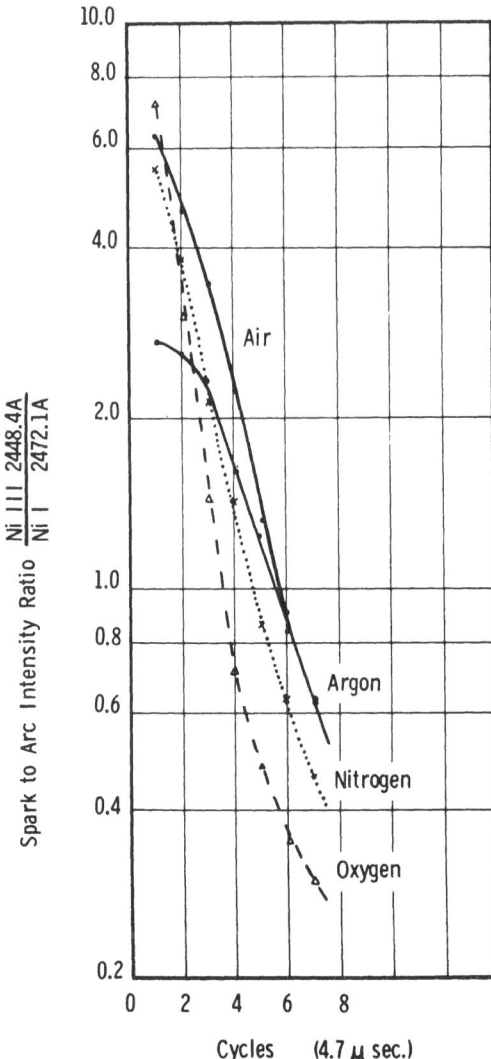

Fig. 9. Variations in the Ni III to Ni I intensity
ratio during the spark for various atmospheres.

This is an expected characteristic since the decay of the spark
transient current is essentially exponential. From what we have previously observed on the line intensities, we might expect some significant differences to appear between the various gases. There are
indeed differences. The effect of the atmosphere is shown in some

radical changes in the spectral character during the lifetime of the spark. Not only does each gas impart its own initial effect, but the atmosphere can be characterized by the rate at which the III : I ratio decays. Whereas oxygen initially imparts the most sparklike character to the spectrum, the rate of decay is the most rapid and ultimately the spectrum ends up with the most arclike character. Argon, on the other hand, is exactly opposite in its effect. It originally gives the most arclike character to the spectrum, decays most slowly, and ends up with the most sparklike contribution.

We have indicated a number of interesting aspects of microsecond-resolution spectroscopy. There are more problems than solutions at this early stage, but the problems are extremely interesting. Further studies of the effect of atmosphere on such simple things as line intensities, intensity ratios, and the spark-to-arc ratio on the microsecond time scale will eventually lead to a better understanding of the processes which influence spectrographic analysis.

## REFERENCES

1. A. Schuster and G. Hemsalech, *Phil. Trans.* **193**, 198, 1900.
2. A. Bardocz, *Spectrochim. Acta* **9**, 307, 1955.
3. A. Bardocz, *Applied Spec.* **11**, 167, 1957.
4. J. W. Beams, Jr., and F. L. Brown, *J. Opt. Soc. Am.* **11**, 11, 1925.
5. S. Smith, *Astrophys. J.* **61**, 186, 1925.
6. E. O. Lawrence and F. G. Dunnington, *Phys. Rev.* **35**, 396, 1930.
7. H. V. Knorr, *Phys. Rev.* **37**, 1611, 1931.
8. L. Blitzer and W. M. Cady, *J. Opt. Soc. Am.* **41**, 440, 1951.
9. J. M. Crosswhite, D. W. Steinhaus, and G. H. Dieke, *J. Opt. Soc. Am.* **41**, 299, 1951.
10. V. A. Fassel and R. W. Tabeling, *Spectrochim. Acta* **8**, 20, 1956.
11. L. B. Loeb and J. M. Meek, *The Mechanism of the Electric Spark*, Stanford University Press, 1941.
12. J. A. Rich, *Resistance Heating in the Arc Cathode Spot Zone*, Paper given at the 13th Annual Gaseous Electronics Conference, 1960.
13. J. M. Somerville, *The Electric Arc*, Methuen & Co., Ltd., London, 1959.
14. T. P. Schreiber and R. F. Majkowski, *Spectrochim. Acta* **12**, 991, 1959.
15. T. P. Schreiber and D. L. Fry, *Spectrochim. Acta* **13**, 99, 1958.
16. L. Minnhagen and L. Stigmark, *Arkiv for Fysik* **13**, 27, 1957.

# Factors Influencing Sensitivity in Atomic Absorption Spectroscopy

Raymond W. Tabeling and John J. Devaney

Jarell-Ash Company
Newtonville, Massachusetts

**The several components comprising an atomic absorption instrument are well known to emission spectroscopists. Nevertheless, the relative merits of these components and their influence upon the sensitivity of their interrelationship have not yet been well defined. The relative noise contributed by the hollow cathode, flame, and measuring system has been investigated and will be reported in this paper. Methods for improving analytical sensitivity will be illustrated by typical examples.**

Since the rediscovery of atomic absorption by Walsh [1] in 1955, much work has been published describing the various applications of this method to chemical analyses. Although various sensitivity limits have been quoted, little has been said concerning the methods used to achieve these sensitivities. It is the purpose of this investigation to define the optimum operating conditions and equipment to achieve maximum sensitivity.

The apparatus used in atomic absorption spectroscopy has been described in detail by others, but because of the wide choice of components available a brief description of the particular system used in this work will be given here. A detailed discussion of each component will follow. The general arrangement of the apparatus is shown in Fig. 1.

The hollow-cathode light source emits the line spectrum of the element to be determined. The sample to be investigated is introduced into the burner, which atomizes the solution and disassociates the molecules into atoms. Relatively few of these atoms will be excited to higher energy levels by the thermal energy of the flame,

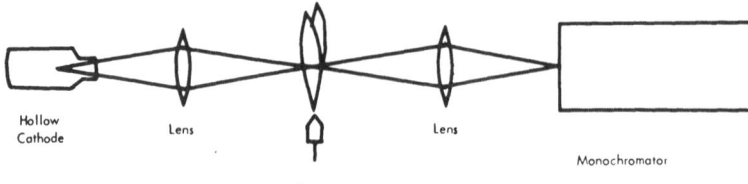

Fig. 1. Diagram of apparatus.

and the great majority of the atoms will be in the ground state and capable of absorbing the light emitted by the hollow cathode. A measurement of this absorption is the means for the definition of the concentration of the element present in the solution. The monochromator isolates the desired line from all others in the spectrum and the intensity of this line is detected and measured by a phototube placed at the exit slit of the monochromator.

Qualitative analysis is thus made possible by the application of the Beer-Bouguer law

$$A = \log_{10}(I_0/I) = abc$$

where $A$ is the absorbance, $I_0$ is the energy incident on the flame, $I$ is the energy transmitted by the flame, $a$ is a constant for any given system, $b$ is the burner length, and $c$ is concentration.

## HOLLOW CATHODE AND POWER SUPPLY

The hollow-cathode discharge tube has been generally accepted as the most suitable light source for atomic absorption. It is now readily available commercially with a wide variety of cathode materials. Quartz window tubes are available for use in the ultraviolet region. These tubes produce very sharp line spectra of high intensity and are very stable even when operated by an unregulated power supply. As pointed out by Walsh [1], sharp lines are necessary in order to obtain measurements of peak absorption.

Great importance should be placed on the hollow cathode's intensity, for several advantages accrue from the use of high-intensity hollow cathode tubes.

In order to point out the importance of the hollow cathode's intensity, Beer's law can be rewritten in terms of what is actually measured:

$$\log I = \log I_0 - abc$$

It is apparent that any increase in $I_0$ causes a corresponding increase in $I$. Since $I$ is the quantity to be measured, it should be as large as possible to eliminate any measuring errors. This can be accomplished by ensuring that $I_0$ is large enough to make the errors introduced negligible.

One of the largest errors introduced is that caused by emission from the flame. Although this can be eliminated by the chopping technique described by Russell, Shelton, and Walsh [2], it can also be eliminated in many cases by making $I_0$ large enough so that the emission intensity from the flame becomes insignificant. Table I shows the emission intensity ($I_{emm}$) for 100 ppm Cu to be 0.1 $\mu$a (photomultiplier output) in this particular experimental setup. If the hollow cathode ($HC$) is run at 6 ma, the total intensity $I_0$ is 0.3 $\mu$a. In this case $I_{emm}$ is 33% of $I_0$. If, however, the hollow-cathode current is raised to 14 ma, $I_0$ becomes 10 $\mu$a, and since $I_{emm}$ remains constant for a given concentration, it is now only 1% of $I_0$. The effect on the measured quantity $I$ can be seen by comparing the 43% absorption measured when $I_0 = 0.3$ $\mu$a with the 45% when $I_0 = 10$ $\mu$a. Thus, the need for chopping can be eliminated in many cases, because at very low concentrations $I_{emm}$ can be considered so small when compared to $I_0$ as to be completely negligible. This has great practical importance because it permits a simplified optical and detection system to be used; in fact, the same system can be used for flame spectroscopy and immediately changed over to absorption by the addition of a hollow-cathode lamp. In practice it will be found practicable for all elements other than the alkalis when high-intensity hollow-cathode lamps are available.

While simple filter photometers have high optical speed, in general the optical speed of an instrument varies inversely with its

**TABLE I**

| $I_0$ $\mu$a | $I_{emm}$, $\mu$a | $I$,% | $HC$, ma |
|---|---|---|---|
| 0.3 | 0.102 | 43 | 6 |
| 1 | 0.1 | 58 | 7 |
| 3 | 0.105 | 64 | 10 |
| 10 | 0.1 | 65 | 14 |

100 ppm Cu 3247 A

bandpass. While filters and low-resolution monochromators are adequate for elements of simple spectra, more complex spectra require higher resolution. If the flame is to be multipassed as is described later, each reflecting and each transmitting surface attenuates the light from 5 to 20%, so that after several passes 50% of the signal can easily be lost. Since photomultiplier detectors, the usual light transducer, develop considerable noise as they are required to deliver more and more gain, the more light available, the better the signal-to-noise ratio in the detector can be, and the better the resulting sensitivity.

It must be cautioned, however, that high hollow-cathode-tube currents can broaden the emission lines from these sources, which results in an effective decrease in absorption, reducing sensitivity. Clearly, these interrelated factors must be balanced to effect the best set of operating conditions for any particular setup.

Since hollow-cathode lamps are gas-discharge devices operated on a voltage plateau, there is little point in voltage-regulating their supply. However, since their voltage-current characteristics are temperature dependent, the most stable electrical and light-output characteristics are obtained by regulating the current supplied to the tube. Hence, in this work the hollow cathodes were excited by a Jarell-Ash #82-135 current-regulated hollow-cathode power supply which is continuously variable from 5 to 100 ma.

Regulation is accomplished by sensing the voltage drop across a resistor in series with the hollow-cathode tube, comparing this to a stable reference voltage, and amplifying the difference which is used to control the output of the supply. Current regulation accomplished by this technique is better than 0.1%. Even with a single-channel system after $1/2$ hr warmup, stability is within 2%/hr.

## BURNER-ATOMIZER

The first burner tried was similar to the one described by Walsh and others. It consisted of a discharge-type atomizer coupled with a burner to give a thin flame about 10 cm long. Many modifications were made in an attempt to achieve a stable flame, but even the best that could be achieved was far inferior to that obtained by Clinton [3]. The discharge-type atomizer used with this burner converted only a small portion of the total sample into very fine droplets which were then fed to the burner. The greater part of the sample was drained from the atomizer by means of a waste tube. Absorption measurements made with this burner showed that it lacked sensitivity. This was attributed partly to the noise, but mainly to the small amount of sample being introduced. Consequently, we went over to an integral-aspirator burner such as that made by Beckman. This type of burner injects the total amount of sample aspirated directly into the burner, thus providing a large concentration of atoms in the flame.

It is known from flame-emission spectroscopy that maximum intensity is found only in certain portions of the flame. In order to find the best region for absorption, the light from the hollow cathode was passed through various regions of the flame and the amount of absorption recorded. Figure 2 shows the amount of absorption plotted as a function of the height above the burner tip.

It is seen that although some absorption occurs in every part of the flame, the best region is 3-4 in. from the burner tip. For maximum sensitivity, a reducing flame approximately 6 in high, and very rich in hydrogen was used. It is also found that the amount of absorption varies from burner to burner and it is worthwhile to investigate a number of burners in order to determine which are the best. Using one burner as described above, sensitivities in the order of 0.5 ppm for Ni and Cu and 2 ppm for Fe were attained. These are considered quite good in that they compare with the sensitivities of the 10-cm burner, although the absorbing path of this flame is only about 2 cm. It is known that sensitivity is approximately proportional to the length of the flame, but it also depends on the number of absorbing atoms in a given flame length. It appears from the above that the 2-cm burner length provided by the Beckman burner is approximately equivalent to the 10-cm burner in common use.

The burner as used in most of our applications was operated at 15 psi oxygen or air. The hydrogen pressure was adjusted to give maximum absorption. Under these conditions the burner consumed about 1.5 cc/min of sample and the flame was about 6 in high.

Because of the small size of the burner it was decided to increase the effective flame length by arranging a number of these burners along the optical axis. This was done in combinations of up to five burners. With all five burners running, the fuel and sample consumption was quite high, and the results obtained indicated that there was little to be gained by the use of more than three burners. Figure 3 shows the increased absorption gained by the use of two additional

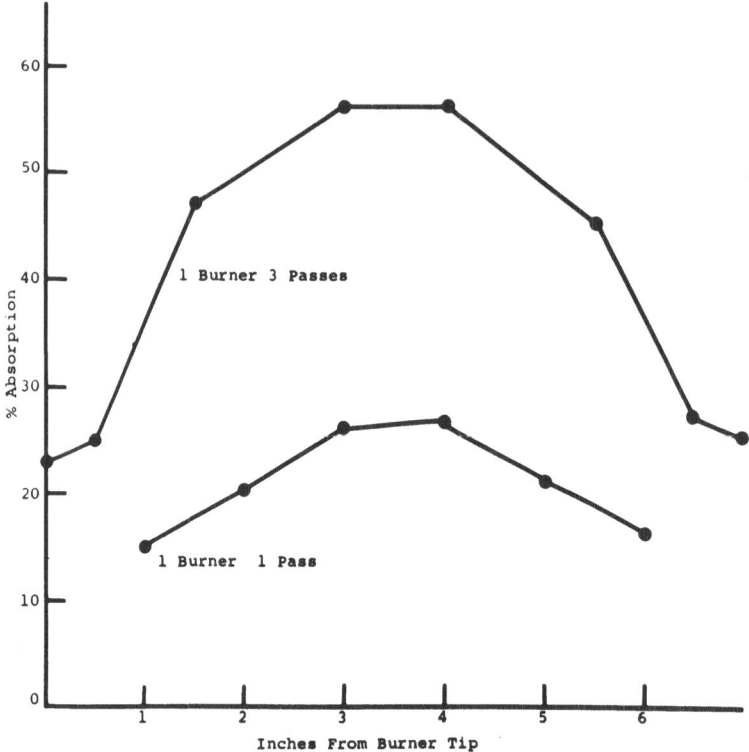

Fig. 2. Absorption plotted as a function of the height above the burner tip.

burners. Three burners in series gave an effective path length of about 6 cm and increased the sensitivity so that 0.1 ppm Ni could be detected.

   Another way to increase the effective flame length is to use a system of multiple passes. This was done as shown in Fig. 4. This should increase the effective length of the flame by a factor of five. Actually, it is not quite that good as all of the passes do not go through the most absorbing regions of the flame. In spite of this, a substantial gain was realized as the sensitivity was increased to 0.05 ppm Ni. A system using seven passes was also tried but the re- flection losses from two additional mirrors required the amplifier gain to be turned up to a point where noise became the limiting factor.

   A J. U. White multipassing system [4] was also tried ex- perimentally, but since it was not practical to use more than about

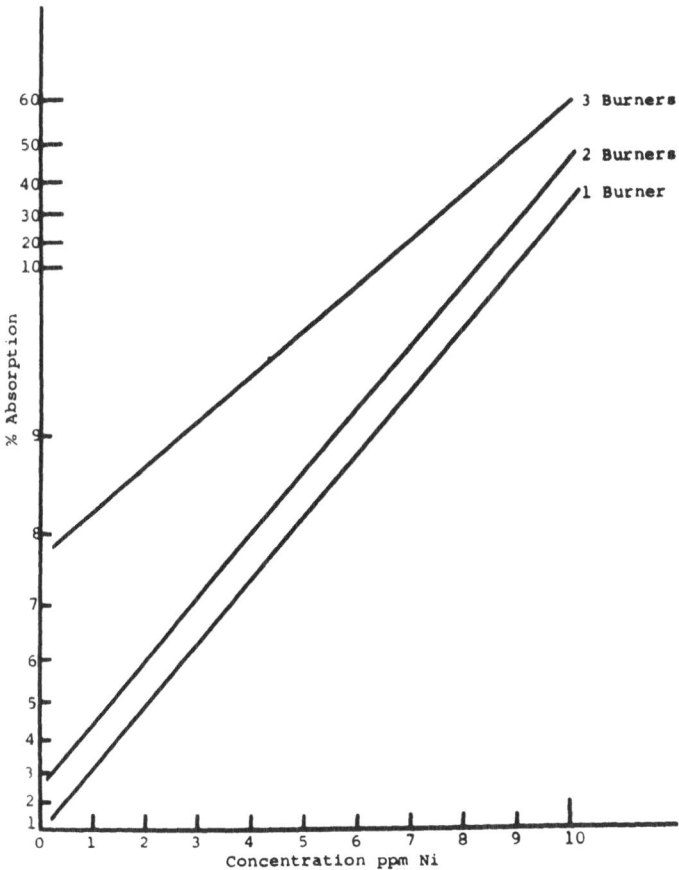

Fig. 3. Increase in absorption with additional burners.

5 passes because of reflection losses, its main advantage, that of providing the possibility of using a large number of passes, could not be used.

The present system was chosen because it not only permitted on-axis admission and exit of the light, which is convenient in optical bar-type mountings, but also offers simplicity of alignment.

## LINE CHOICE

Another factor involving sensitivity, and probably the most important, is the line chosen. It has been pointed out by others that

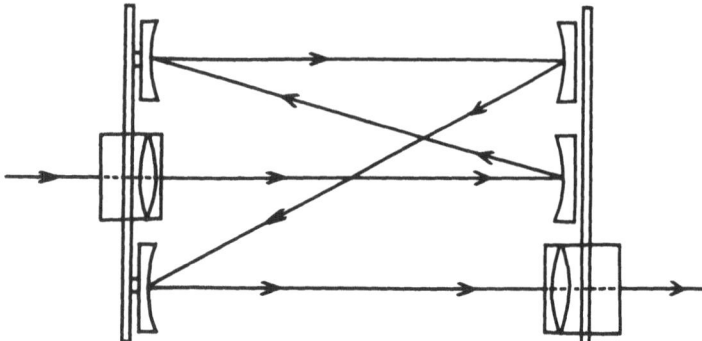

Fig. 4. Multiple-pass system.

lines which return to the ground state are the best lines for absorption. Unfortunately, there is no way to predict which line is the best. An experimental method for finding the best line has been described by Allan [5], and a somewhat similar method has been used in this laboratory. Tables prepared by Charlotte Moore [6] list lines of the elements and their associated energy levels. After listing all the lines which return to 0.00 ev, one then scans the monochromator over the spectrum and examines each line for percent absorption. Table II shows a list prepared for copper. A 100-ppm Cu sample was used and the amount of absorption for each line was recorded. This shows the Cu 3247 A line to be the best.

### TABLE II
### Absorption of Various Cu Lines

| Line, A | Excitation potential, ev | | Intensity | Absorption, % |
|---------|-----|------|-----------|---------------|
|         | Low | High |           |               |
| 2165.1  | 0.00 | 5.70 | 1300R | 25 |
| 2178.9  | 0.00 | 5.66 | 1600R | 30 |
| 2181.7  | 0.00 | 5.66 | 1700R | 30 |
| 2225.7  | 0.00 | 5.54 | 2100R | 16 |
| 2244.3  | 0.00 | 5.50 | 2300R | 0 |
| 2441.6  | 0.00 | 5.05 | 1000R | 0 |
| 2492.1  | 0.00 | 4.95 | 2000R | 3 |
| 3247    | 0.00 | 3.80 | 1000R | 85 |
| 3247    | 0.00 | 3.77 | 600R | 62 |

Sample, 100 ppm Cu; 1 burner

The fact that different lines have different sensitivities can be put to good use. By looking at Fig. 5 it can be seen that the Cu 3247 A line is the most sensitive, but it cannot be used above 100 ppm. In order to accomplish this, the Cu 2244 A line can be used as it is quite linear in the 100- to 10,000-ppm range. Other less sensitive lines could also be used for higher concentrations.

### MONOCHROMATOR

Because of the complex spectra of Fe and Ni, a good monochromator is necessary to isolate the desired line from all others. The instrument used in this work was a Jarrell-Ash Model 82-000 Ebert Scanning Spectrometer. When used with a 30,000-line/in. grating, this unit gives a linear dispersion of 16 A/mm and resolution of at least 0.2 A in the first order. This resolution is required to separate the Ni 2320.03 A line from the one at 2319.76 A. If this is not done, the intensity of the 2319.76 A line is included in $I_0$. This has the effect of broadening the absorbing line and diminishing its sensitivity. Figure 6 shows the complexity of the Ni spectrum in this region.

Another advantage of this equipment is the fact that by merely

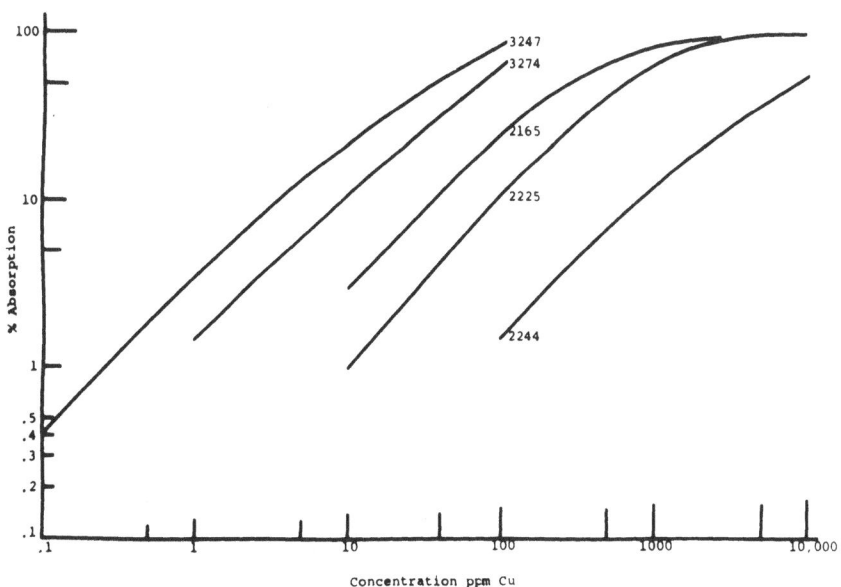

Fig. 5. Copper—variation of sensitivity using various lines.

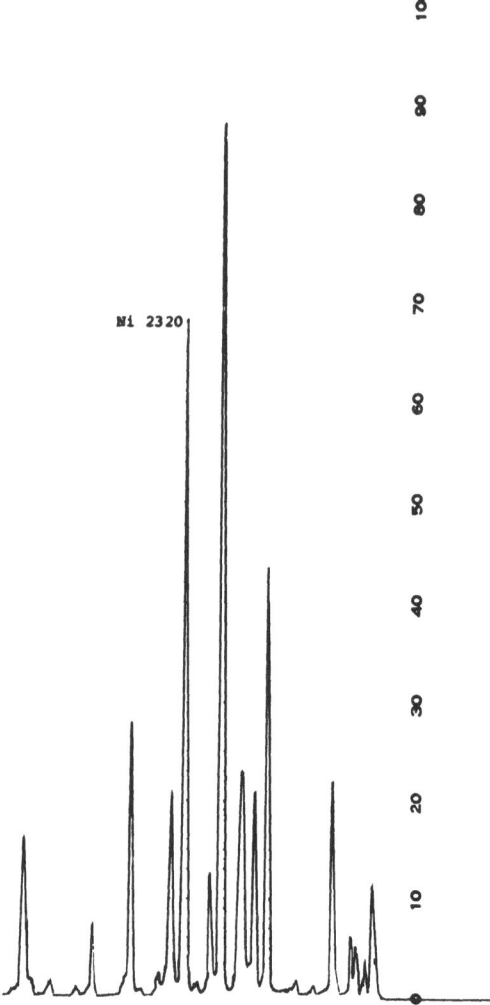

Fig. 6. Nickel spectrum.

turning off the hollow cathode, one has at his disposal an excellent flame photometer.

It has been shown how each component, when considered separately, contributes to the sensitivity of the complete instrument. In order to appreciate the sensitivities achieved by the proper utilization of each component, atomic absorption spectroscopy should be compared with other methods. Table III shows the sensitivity limits for two of the most common methods of chemical analysis [6]. Under atomic absorption there are two columns. Column A lists the best

## TABLE III
### Sensitivity Limits (ppm) for Three Analytical Methods

| Element | Spectrographic | Flame | Atomic absorption | |
|---|---|---|---|---|
| | | | A | B |
| Fe | 0.5 | 6 | 2.5 | 0.1 |
| Ni | 4 | 1.0 | 1.0 | 0.01 |
| Cu | 0.05 | 0.6 | 0.5 | 0.01 |

sensitivities achieved at this laboratory. The comparison is made to illustrate the substantial increase in sensitivity that can be made by the optimum choice and utilization of components.

## REFERENCES

1. A. Walsh, *Spectrochimica Acta* **7**, 108, 1955.
2. B. J. Russell, J. P. Shelton, and A. Walsh, *Spectrochimica Acta* **8**, 317, 1957.
3. O. E. Clinton, *Spectrochimica Acta* **16**, 985, 1960.
4. J. U. White, *J. Opt. Soc. Am.* **32**, 285, 1942.
5. J. E. Allan, *Spectrochimica Acta* **10**, 800, 1959.
6. C. Moore, *A Multiplet Table of Astrophysical Interest*, Princeton University.
7. J. W. Robinson, *Anal. Chem.* **32**, 17A, 1960.

# Analysis of Europium Oxide for Rare Earth Elements and Other Impurities

J. T. Rozsa and J. Stone

National Spectrographic Laboratories, Inc.
Cleveland, Ohio

Quantitative spectrographic procedures have been applied to the analysis of europium oxide. Because of its relative volatility in comparison to the other rare earth elements, the total-energy and carrier-method approach had to be modified to minimize background. Some 20 elements are included in the range of 1 to 10,000 ppm. Precision has been found to be 5 to 20%.

The analysis of purified lanthanide elements poses several very interesting technical problems. These elements, while chemically similar, have widely different evolution characteristics, spectrographically [1,2]. Europium is the most volatile of the group, and in addition, has very complex spectra. A literature survey does not reveal very much specific information on the analysis of purified europium oxide. The extensive work of Fassel [3] and his collaborators on other members of the lanthanide group was, of course, helpful.

## TOTAL-ENERGY METHOD

The total-energy method first described by Slavin [4] was the initial subject of study. Extensive previous experience with rare earth analysis in this laboratory had demonstrated that the total-energy method would offer the optimum possibility for precision and modest sensitivity.

To control the unusually high volatility of europium without suppression of the impurity elements, a mixture of carbon black and –325 mesh carbon was mixed intimately with the sample. The rapid formation of the carbide bead within 5 sec, as indicated by x-ray

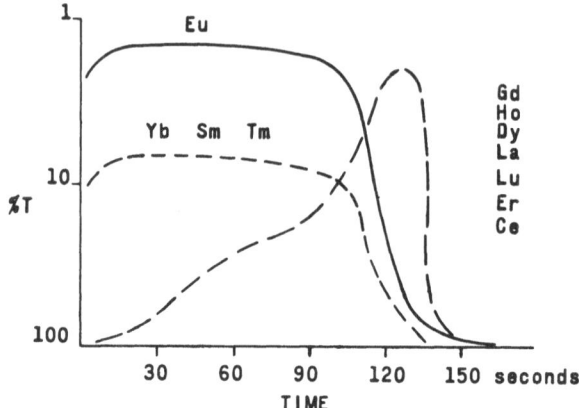

Fig. 1. Total-energy excitation

diffraction analysis, permitted a reproducible controlled evolution of impurities into the arc column.

An arc programing procedure was utilized for excitation in which the arc amperage was held to the low value of 10 amp for 15 sec and then raised to 30 amp for 120 sec, without interruption of the arc. Figure 1 illustrates the typical time of wait obtained. The graph is somewhat condensed for ease of presentation. In addition to providing the few seconds for complete material carburization, the initial low amperage also materially reduced spattering of the charge from the electrode cup and promoted obtainment of 45° working curves.

In programing the arc, various amperage values were investigated but the greatest improvement in sensitivity was realized for the 10- to 30-amp regime, particularly for the titanium and zirconium determinations. A comparative chart, Table. I, partially reveals the influence of arc programing upon sensitivity. The inert-gas combination of argon and oxygen was helpful but the improvement did not warrant the additional precautions at this time.

## FRACTIONAL-DISTILLATION METHOD

To obtain the requisite sensitivity, the carrier distillation method of Scribner [5] was utilized. Using this method evoked some misgivings since no carbon buffer is normally used in a carrier distillation method. To suppress the europium sufficiently long to permit evaluation of impurities, gallium oxide alone has proved satisfactory.

Figure 2 illustrates, by a time-of-wait curve, why silver chloride and several others were unsatisfactory.

A particular effort was made to develop overlapping working curves by the total-energy method as a check. This system tends to ensure that any irregular and unexpected matrix effect will be obvious. Reliance is placed upon the total-energy method.

## INSTRUMENTAL DETAILS

The method summary is detailed in Table II. The analytical line pairings are listed in Table III. The wavelengths follow the tables prepared by J. A. Norris [6].

A high-dispersion spectrograph, Bausch and Lomb Dual Grating, is utilized to handle the complex spectrum. Copper, titanium, and zirconium required 1.2 A/mm reciprocal linear dispersion to eliminate interferences.

The alkali method is a total-energy type since lithium, with its erratic refractory nature, makes the use of carrier methods precarious. No standards are available, hence it was necessary to synthesize the reference samples. Weighed C. P. oxides of the respective metals are carefully hand ground under alcohol into a repurified europium oxide to the desired impurity concentrations. Residual element concentrations in the blank or base oxide are determined and values corrected by the Pierce approximation method [7].

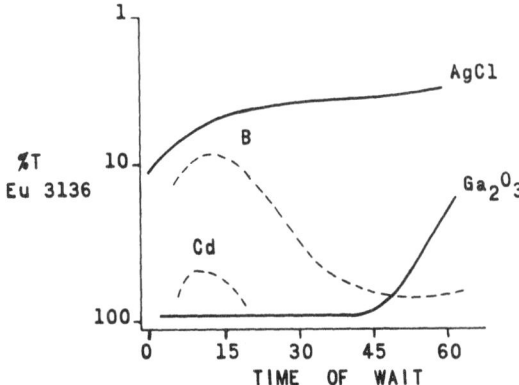

Fig. 2. Carrier distillation evolution for $Eu_2O_3$.

## TABLE I
### Effect of Excitation Form upon Sensitivity*

| Element | Method and Excitation | | |
|---|---|---|---|
| | Total-energy: 30-amp ac arc | Total-energy: 10- to 30-amp programed arc | Carrier: 20-amp dc arc |
| Sm | 100 ppm | 20 ppm | 2 ppm |
| Gd | 10 | 5 | 100 |
| Pb | 100 | 11 | 1 |
| Si | 15 | 2 | 14 |
| Zr | 14 | 7 | 50 |
| B | 1 | 1 | 2 |

*Entries represent typical sensitivities.

## TABLE II
### Summary of Spectrographic Method for Analysis of $Eu_2O_3$

| | Method | | | |
|---|---|---|---|---|
| | Total-energy | | Carrier | Total-energy |
| Excitation | Sustaining | | dc arc | Sustaining |
| Amperage | 10–30 | | 20 | 18 |
| Exposure, total | 15–120 | | 50 | 90 |
| Spectrograph | Dual | | Dual | Dual |
| Range, A    A | 2700–3200 | 3250–3750 | 2500 – 2500 | 5850 – 6850 |
|          B | 3260–3560 | 3900–4400 | 2100 – 2600 | 7000 – 8000 |
| Dispersion,   A | 2 | 2 | 4 | 4 |
| A/mm      B | 1.2 | 2 | 2 | 4 |
| Filters,     A | 12 | 12 | 100 | 100 |
| %T        B | 100 | 100 | 100 | 100 |
|    Over-all | 100/12 | 100/12 | 100/12 | 100/12 Corning |
| Electrode | | | | |
| Charge, mg | 35 | 35 | 50 | 15 |
| Diluent | C 50% | C 50% | $Ga_2O_3$ 10% | C 50% |
| Electrodes | L 3751 | L 3751 | L 4036 | L 3751 |
| | L 3709 | L 3709 | L 3918 | L 3777 |

## TABLE III
### Analytical Line Pairings for Analysis of $Eu_2O_3$

| Method | Element | Wavelength, A | Range, ppm |
|---|---|---|---|
| Total-energy | Eu | 3137 | Int. Stnd. |
| | Mg | 2791 | 1– 100 |
| | Mn | 2801 | 1– 250 |
| | Pb | 2802 | 10– 700 |
| | Cr | 2835 | 3– 400 |
| | Sn | 2840 | 10– 300 |
| | Hf | 2820 | 100–1000 |
| | Si | 2881 | 2– 200 |
| | Th | 2982 | 100–1000 |
| | Ni | 3002 | 2– 200 |
| | Fe | 3020 | 4– 800 |
| | Al | 3082 | 10– 200 |
| | Mo | 3133 | 5– 200 |
| | Cu | 3274 | 10– 100 |
| | Ti | 3349 | 10– 100 |
| | Zr | 3392 | 10– 100 |
| | Na | 5890 | 1– 100 |
| | Ba | 6142 | 1– 85 |
| | Li | 6708 | 1– 100 |
| | K | 7665 | 1– 100 |
| Carrier | – | 50%T | Int. Stnd. |
| | Cd | 2288 | 0.8 – 10 |
| | Co | 3453 | 1– 10 |
| | B | 2498 | 1– 150 |
| | Si | 2881 | 1– 30 |
| | Cu | 3274 | 1– 30 |
| | Ag | 3282 | 1– 10 |
| | Zn | 3345 | 10– 110 |
| | Ca | 4227 | 1– 100 |
| | Cr | 4254 | 1– 100 |
| | Sm | 4281 | 2– 50 |
| | Yb | 3694 | 5– 50 |

## TABLE III (continued)

| Method | Element | Wavelength, A | Range, ppm |
|--------|---------|---------------|------------|
| Total-energy | Eu | 3391 | Int. Stnd. |
| | Gd | 3422 | 5– 300 |
| | Sm | 3628 | 20– 500 |
| | Dy | 3385 | 1– 400 |
| | Er | 3373 | 1– 500 |
| | Tm | 3462 | 10– 500 |
| | Yb | 3694 | 1– 400 |
| | Sc | 3630 | 1– 300 |
| | Ho | 3400 | 10– 300 |
| | Ce | 4012 | 75– 500 |
| | La | 3988 | 5–1000 |
| | Y | 3629 | 9– 100 |

## TABLE IV
### Precision Data for Analysis of $Eu_2O_3$

| Method | Element | Concentration, ppm | Coefficient of variation |
|--------|---------|--------------------|--------------------------|
| Total-energy | Yb | 52 | 8.9 |
| | Pb | 101 | 13.5 |
| | Mn | 100 | 13.0 |
| | Si | 110 | 5.6 |
| | Fe | 100 | 6.5 |
| | Tm | 52 | 6.6 |
| | Y | 52 | 5.8 |
| | Gd | 52 | 5.3 |
| | Ti | 104 | 7.6 |
| Carrier | Cd | 100 | 19.2 |
| | Si | 52 | 17.6 |

# PRECISION

Repetitive precision studies indicate that a normal coefficient of variation of 6 to 10% is obtained for the total-energy method, and 20% is obtained for the carrier method. (See Table IV.)

No comparable methods were available for accuracy study, but metallurgical applications have confirmed numerous analyses.

# CONCLUSION

A method has been developed for the spectrographic analysis of 37 elements, including 13 rare earth elements in europium oxide. The concentration range of the impurities is 1 to 1000 ppm with a precision of 10 or 20% coefficient of variation.

# REFERENCES

1. J. T. Rozsa, J. Stone, and D. C. Manning, "Emission Spectrographic Determination of Rare Earths and Other Elements in Yttrium Oxide," *Pittsburgh Conference on Analytical Chemistry*, March 3, 1961.
2. F. Trombe, *Bull. Soc. Chem. France* **10**, 1010, 1953.
3. V. A. Fassel and H. A. Wilhelm, *J. Opt. Soc. Am.* **38**, 6, 518, 1948.
4. M. L. Slavin, *Proc. Sixth Summer Conference on Spectroscopy*, John Wiley and Sons, 1939.
5. B. F. Scribner and H. R. Mullin, *J. Res. Natl. Bur. Stnds.* **37**, 379, 1946.
6. J. A. Norris, *Wavelength Tables of Rare Earth Elements ORNL* 2774, 1960.
7. W. C. Pierce and N. H. Nachtrieb, *Ind. and Eng. Chem., Anal. Ed.* **13**, 774 1941.

# Progress in Atomic Absorption Spectroscopy

J. W. Robinson

Esso Research Laboratory
Baton Rouge, La.

The field of atomic absorption, including advantages and disadvantages, variables, elements determinable, forced-feed burners, spark sources, flame adapters, and dispersion requirements is reviewed. A comparison is made with flame photometry.

## ADVANTAGES AND DISADVANTAGES

The principal advantages of atomic absorption spectroscopy, based on experimental work, include a high degree of sensitivity, freedom from interference from other metals present, and simplicity of operation.

The principal disadvantage seems to be that when a flame is used as the atomizer, a number of metals form a refractory oxide and are, therefore, not detectable in this system.

In order to obtain reproducible analytical results, it is necessary to control certain variables. These are similar to the variables encountered in flame photometry and include: (1) rate of sample feed into the burner, (2) type of fuel used in the flame, (3) ratio of fuel to oxygen in the flame, (4) the part of the flame examined, (5) the solvent used in the sample, and (6) the ion or organic addend with which the sample metal is associated in solution.

It will be noted that each one of these variables directly affects the number of atoms produced in the flame and, therefore, the number of atoms in the light path.

## RECENT ADVANCES

### Forced-Feed Burner

A burner has been developed in which the sample was mechanically nebulized and sprayed into a flame for reduction to atoms.

The feed rate was independent of the flow rate of the fuel and oxygen and was, therefore, independent of aspiration. Results indicated that the optimum conditions for absorption and emission are quite different; further, these conditions are extensively modified when an organic solvent is used instead of an aqueous solvent. Results are listed in Table I.

TABLE I
**Optimum Conditions for Emission
and Absorption, Using A Forced-Feed Burner**

| Solution | Fuel pressure, psi | | | |
|----------|---------|---------|---------|---------|
|          | Atomic absorption | | Emission | |
|          | $H_2$ | $O_2$ | $H_2$ | $O_2$ |
| Aqueous  | 2.0 | 0.6 | 3.5 | 8.0 |
| Organic  | 0.2 | 0.6 | 0.2 | 2.0 |

## Effect of Solvent

Solutions of nickel naphthanate were made up in various organic solvents and a relative degree of absorption at fixed feed rates was measured. It was found that in the solvents studied the degree of absorption was almost independent of the particular solvent used. This is in direct contrast to results obtained when an aspiration burner is used. These results indicate that the atomic absorption may be free of interference not only from other metals, but from the matrix of the sample.

## Flame Profile

The intensity of emission and the relative degree of absorption were measured at different parts of the flame. The results show that the maximum absorption took place at a considerably higher part of the flame than the point of maximum emission. This was probably because the lifetime of an excited atom is much shorter than the lifetime of an unexcited atom, thus allowing some accumulation of ground-state atoms as compared to excited atoms. The results thus indicate that the excited atoms are not in direct thermal equilibrium with the ground-state atoms, otherwise the points of maxima should nearly coincide, some difference being obtained because of change in flame temperature with the increasing height. However, these results seem to indicate that at the point of maximum emission, extra excitation is caused by the other energy sources such as chemiluminescence or ultraviolet light.

## Atomization Using Electrical Discharge

In the past, a flame has been used principally for the atomization of the sample. One of the more serious disadvantages of the use of a flame was that some metals formed refractory oxides during combustion and were not reduced to the metallic state. This prevented their detection and determination by atomic absorption spectroscopy. It was anticipated that if these metals could be broken down to the atomic state by an alternate means, they would be determinable. The metal studied was aluminum. Attempts to reduce this to the atomic state in flames have been unsuccessful, even using oxycyanogen flames at flame temperature up to 4500°C.

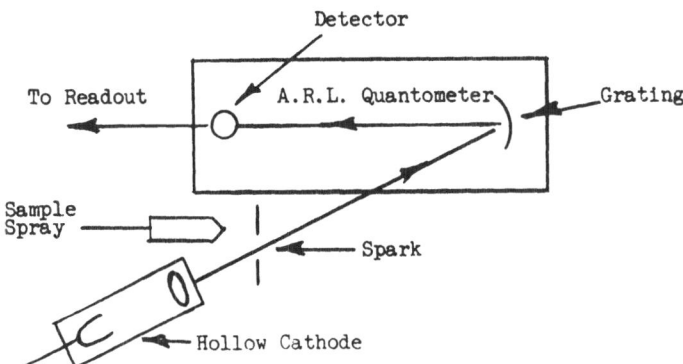

Fig. 1. Optical alignment for spark and spray atomization of aluminum.

An ARL Quantometer was used as a spark detection system purely for demonstration purposes. The optical alignment is shown in Fig. 1. The sample was sprayed into the discharging spark, using an insulated forced-feed burner. When the electrode was discharged across a flame containing the sample, no visible discharge occurred and no absorption by the metal was detected. However, when the metal was merely sprayed into the electrical discharge, appreciable absorption took place. With results obtained at a wavelength of 3944 A, the limit of detection of aluminum was 1 ppm. This illustrated the high efficiency of this system in producing unexcited metals in the atomic state.

## Flame Adapter

The extent of absorption by metal atoms is a direct function of the number of unexcited atoms in the light path. In an attempt to increase this number, an adapter was designed to change a narrow flame into a broad flame. It was hoped that the number of unexcited atoms in the light path would be increased and the number of excited atoms in the light path would be decreased. This should lead to an increase in the absorption signal and a decrease in the emission signal from the flame. The flame adapter is illustrated in Fig. 2.

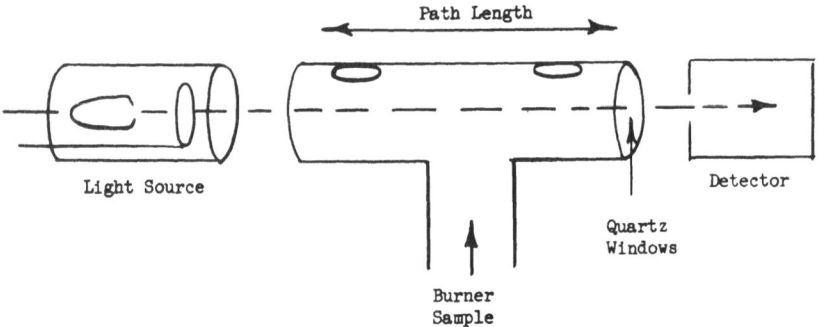

Fig. 2. Flame adapter, schematic.

The flame used was an oxyhydrogen flame and the samples were in aqueous solution. In preliminary tests using sodium and nickel, however, the adapter was not successful. In both cases, presumably, the atoms were oxidized before reaching the pertinent part of the adapter. However, when an aqueous solution of platinum was aspirated, a tenfold increase in absorption was observed. The limit of detection reached was about 1 ppm in this case. The results lead us to conclude that the adapter could not be used for the determination of the alkali metals and some of the transition metals; however, it should be useful for determination of low quantities of the noble metals. Further, it is anticipated that if flame conditions could be regulated to prevent the formation of metal oxides from the metal in the upper portions of the flame, its use might be extended to include other elements, particularly the transition elements.

## Dispersion Requirements

It was shown that with decreasing slit widths, the limits of detection for nickel, using a wavelength of 3414 A, were progressively decreased. This was probably because less unabsorbable light was allowed to fall on the photomultiplier detector, thus improving the signal-to-background relationship. Results showed that with a spectral slit width of 3.2 A, the limit of detection of Ni (3414 A) was 5 ppm, but with a spectral slit width of 25 A, the detection limit was 10 ppm.

These results show that for transitional elements, at least, a suitable prism or grating should be used in order to obtain high sensitivity. However, for alkali metals where other lines are somewhat remote from the absorbed lines, a light filter would probably be sufficient.

# Applications of Quartz Prism Spectrophotometers—A Review

R. J. Manning

Beckman Instruments
Lincolnwood, Illinois

A brief review of ultraviolet, visible, and near-infrared spectrophotometry, covering absorption, transmission, emission, and reflectance studies. Uses of the relatively new far-ultraviolet region are covered.

# The Evaporation of Boron from Acid Solutions and Residues

## C. Feldman

Oak Ridge National Laboratory
Oak Ridge, Tennessee

In determining traces of boron it is often necessary to heat or evaporate an initially acid solution without losing boron. This loss can be prevented by introducing excess alkali, but this is often inconvenient and complicates the final spectrographic determination. It was therefore decided to (1) determine the seriousness of losses of boron during the evaporation (or heating) of various acid solutions at steam-bath temperatures ($\sim 80°C$) and above, and (2) look for a method of preventing such losses if they were substantial.

Various acid solutions containing boron and cobalt were sampled at successive stages of evaporation and the boron-cobalt concentration ratio measured. Analytical exposures were made with carbon porous cups on an ARL Quantometer. The intensity ratio B 2497.73 : Co 2582.24 was measured; the mean relative deviation of duplicate exposures made with different electrodes was $\pm 1.88\%$.

Less than 5% of the boron was lost from any of the acid solutions, except HCl, until the volume was very small. Boron volatilized relatively quickly from the dried residues. However, the presence of a fifteenfold molar excess of mannitol prevented this loss when not destroyed by the acid.

Experiments indicated that the fuming of $H_2SO_4$ and $HClO_4$ and/or the subsequent baking of residues so obtained caused substantial losses of boron. The accuracy of procedures involving such steps is thus dependent on making the fractional loss of boron equal in samples and standards.

# Infrared and Raman Spectroscopy
## and
## Gas Chromatography

# Sample Problem in Raman Spectroscopy*

## Marvin C. Tobin

Chemical Research Center
American Cyanamid Co.
Stamford, Conn.

---

The sample problem has inhibited widespread use of Raman spectroscopy. The main aspects of the sample problem are turbidity, color, and fluorescence. Turbid samples may be successfully run by suitable filtering of exciting light and by using a sharp bandstop filter to remove exciting light from the Raman spectrum. Theoretical analysis shows that a Toronto arc source combined with a conical sample tube will give the highest yield of Raman radiation. The efficiencies of alternative sample shapes and optical designs, including beam-splitting arrangements, may readily be calculated by use of general principles.

The problems of color and fluorescence seem insuperable as long as the mercury blue line is used for excitation. The mercury green line and helium yellow line are useful sources, which do not strongly excite fluorescence if short-wavelength light is filtered out. The cadmium red and lithium red lines are other likely long-wavelength source lines. It seems entirely feasible to construct a turbid-sample source using yellow or red exciting light and thus get rid of the sample problem entirely. However, such a source would at present be limited to photographic recording.

Use of laser or other high-intensity sources might change this picture entirely.

Raman spectroscopy has never reached the advanced state of development of infrared spectroscopy. While bench-model infrared spectrometers for use by organic chemists are to be found today in every large industrial laboratory, Raman spectroscopy is carried out mainly in universities or research institutes. The reason for this is not, as is sometimes stated, that infrared spectroscopy provides the same information. Double and triple bonds, S-S bonds, low-

*A Contribution from the Chemical Research Department, Central Research Division, American Cyanamid Company.

lying frequencies and a host of other molecular properties are best studied through Raman spectroscopy. The limited use of Raman spectroscopy, rather, arises from the set of nuisances known collectively as the sample problem.

It will be recalled that a Raman spectrum is excited by illuminating a sample with intense monochromatic radiation. The scattered light contains, in addition to the exciting wavelength, additional wavelengths corresponding to the Raman lines. The intensities of the Raman lines are generally $10^{-2}$ to $10^{-4}$ that of the scattered exciting line, the Rayleigh line. In order to get intense exciting light, it has proved necessary to use line-emission sources, most commonly mercury. One is thus limited to those exciting wavelengths which are naturally strong lines in an emission spectrum. Furthermore, it is necessary to use elements with few emission lines as sources, so that other emission lines should not block Raman lines. Extraneous emission lines, which might also excite the Raman spectrum, must be suppressed by suitable filters.

The most convenient line to use has been the blue Hg 4358 A line. This line is intense, the neighboring lines are readily suppressed, and the mercury lamps are not difficult to use. The use of the Hg 4358 A line brings us to the first aspect of the sample problem, color. If a sample has a strong absorption band in or near the blue region of the spectrum, the exciting line is absorbed, and little or none of it gets in to excite a Raman spectrum. Conventional Raman spectroscopy is, then, limited to samples which are colored no more deeply than a pale yellow. The second aspect of the sample problem, fluorescence, also appears often when the Hg 4358 A line is used. Most samples, as prepared, contain small amounts of fluorescent impurities whose fluorescence completely swamps the Raman lines. It has proved necessary to purify by painstaking sublimation, distillation, or recrystallization in order to get rid of impurities.

The third aspect of the sample problem is turbidity. A turbid sample, such as a crystal powder, reflects much of the exciting radiation back out of the sample. Furthermore, the powder reflects the exciting wavelength so strongly into the spectrograph that the plate is completely blackened. For these reasons, most reported Raman spectra have been obtained on purified colorless liquids.

In spite of the somber picture just painted, substantial progress has been made in the last five years in overcoming the sample problem. In particular, several effective crystal powder sources have been

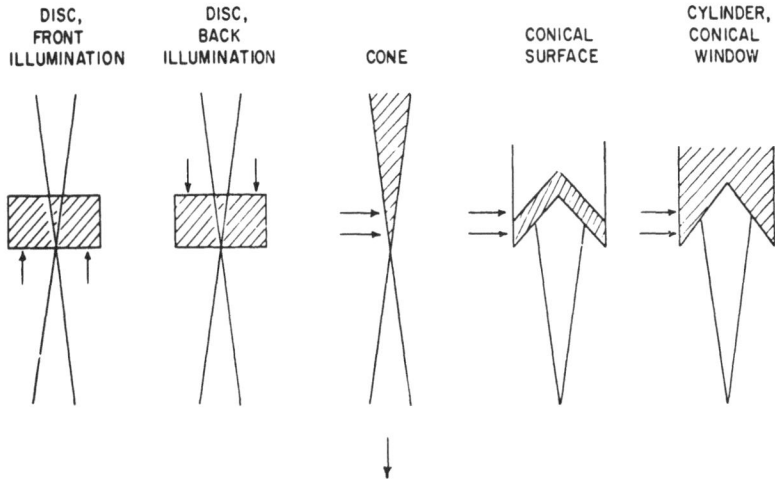

Fig. 1. Methods of illuminating samples for Raman spectroscopy.

designed and a number of useful sources in the yellow and red region of the spectrum have been described. The problem of fluorescence is still troublesome, but there are indications that this too can be ameliorated by using long-wavelength exciting lines.

There are certain features common to the problems of color and turbidity, so we may conveniently start with a discussion of the latter. A turbid sample decreases Raman intensity in two ways.* First, material in inner layers of the sample is shielded from the exciting light by outer layers of the sample. Second, Raman radiation originating in parts of the sample distant from the spectrograph is shielded by parts of the sample near the spectrograph. This situation is made clearer by a consideration of Fig. 1, which shows a number of possible ways of illuminating a Raman sample. Since only those portions of the sample lying in the spectrographic cone of aperture are effective in delivering radiation to the spectrograph, it is necessary to design Raman sample holders with no sample lying between the source and the aperture cone. The fifth tube shown is, therefore, poorly designed. This remark is equally valid for colored liquids, so these

*The discussion of the sample problem in this paper is based on spectrographs with conventional optics. Somewhat different considerations hold for unconventional systems like the Cary Raman Spectrophotometer.

should not be examined in a cylindrical Raman tube. Colorless liquids transmit the exciting light; thus, use of a cylindrical tube does no harm other than increasing the sample size.

There is little that can be done about the intrinsic scatter of a crystalline powder. It is sometimes possible to fuse a powder to a polycrystalline solid or to moisten it with a liquid matching its refractive index. These steps add substantially to the ease of obtaining spectra. Aqueous salt solutions or $CCl_4 - CS_2$ solutions are advantageous moistening liquids. It is possible to calculate the flux delivered to the spectrograph by sample holders of various shapes; this is shown in Fig. 2. A Raman tube matching the cone of aperture delivers about twice the flux of the front- or back-illuminated disk. There are several interesting results of these equations. The first is that for the back-illuminated disk there is an optimum thickness given by $a = 1/\epsilon$ cm. The second is that since transmitted exciting

---

*1. CONE OF APERTURE*

$$F = (2\pi B/\epsilon^2) \int_0^a (e^{-\epsilon x}/x^2) \cdot$$

$$(e^{-\epsilon x \tan a} + \epsilon x \tan a - 1)dx$$

*2 BACK-ILLUMINATED DISC*

$$F_{max} = \pi B \tan^2 a/e\epsilon$$

*3. FRONT-ILLUMINATED DISC*

$$F = (\pi B \tan^2 a/2\epsilon)(1 - e^{-2\epsilon a})$$

*4. EFFICIENCIES FOR $\epsilon = 2$ IN TYPICAL CASE*

1.  $F = 0.0060 B (a = 1 cm)$

    $F = 0.0182 B (a = 1 cm, \epsilon = 0)$

2.  $F_m \doteq 0.0023 B$

3.  $F = 0.0031 B (a = \infty)$

Fig. 2. Flux reaching spectrograph from various Raman sample shapes [1] $B$ = sample brightness, $\epsilon$ = sample specific extinction coefficient, $\alpha$ = spectrograph aperture, $a$ = sample depth.

light goes down as $\exp(-a\epsilon)$, the ratio of scattered flux to transmitted exciting light is proportional to the thickness $a$. We thus have the interesting result that by making our sample sufficiently thick, we may get rid of excessive exciting light. We pay for this, of course, with long exposure times. The gist of the foregoing remarks is that in working with any Raman sample whatsoever, the most efficient tube shape is a Raman tube matched to the spectrographic cone of aperture.

Once we have our Raman radiation from a turbid sample, excessive exciting light must be removed. The best current method of doing this is shown in Fig. 3. The collimated light from the Raman-tube window is reflected from multilayer interference filters which transmit Hg 4358 A and reflect other wavelengths. While the band-stop filter uses two reflectors, experience has shown that introduction of a third would greatly increase efficiency. Use of a bandstop system effectively eliminates the problem of excessive reflection in a turbid sample.

Fig. 3. Bandstop filter unit.

We come next to the question of the exciting source. What is desired is to illuminate the sample with as intense a monochromatic luminous flux as possible.

This has been done in the past in two ways. First, the sample may be surrounded by a Toronto arc, whose output is filtered by suitable chemical media [1]. Second, the output from a mercury arc may be collimated, passed through an interference filter, and focused on the sample [2]. What is really in question is the fraction of the $4\pi$ steradians of solid angle subtended at a point sample which is surrounded by a luminous source. The results of a calculation are shown in Fig. 4. It turns out that a Toronto arc delivers about ten times the

---

*1. TORONTO ARC, SAMPLE AT CENTER*

$$F = 4\pi B'p \tan^{-1}(h/n)$$

$$F = 14.80 \, B' \text{ for Standard Arc}$$

*2. COLLIMATING MONOCHROMATOR*

$$F = 2\pi B'p(\gamma - 1)/\gamma$$

$$\gamma = \sqrt{1 + (1/4f^2)}$$

$$F = 1.33 B' \text{ for f-1 lens}$$

---

Fig. 4. Illumination on a point sample from a Toronto arc and from an $f/1$ collimating monochromator [1]. $B'$ = source brightness per unit depth, $p$ = source depth, $h$, $n$ = Toronto arc dimensions, $f$ = collimating lens aperture.

luminous flux to a sample as does a single collimator filter with an $f/1$ lens. In order to use a Toronto arc, it is, of course, necessary to have a suitable chemical filter to suppress unwanted background. Such a filter has been developed for Hg 4358 A [1].

We may sum up by saying that methods are now available for obtaining Raman spectra of colorless nonfluorescent crystalline powders with convenient apparatus and reasonable exposure times.

Figures 5, 6, and 7 show some original spectra taken with the Cyanamid crystal powder source. It will be noted that the lines are satisfactorily strong, and that background is not excessive. The background in Fig. 5 is believed to be due to fluorescence.

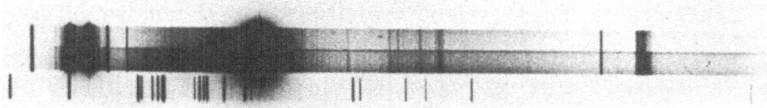

Fig. 5. Raman spectra of Marlex 50 (top) and DYNH (bottom) polyethylenes, 3-hr exposure.

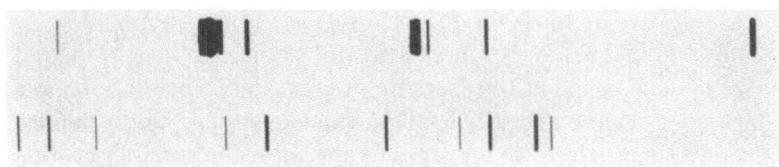

Fig. 6. Raman spectrum of crystalline thiourea, 3-hr exposure.

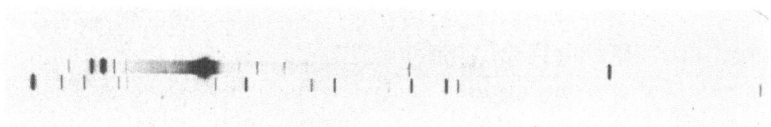

Fig. 7. Raman spectrum of crystalline $Cr(CO)_6$, 4-hr exposure with bandstop unit augmented by Wratten #3387 filter. (Some weak lines at 1000 to 1400 $cm^{-1}$ not visible in this figure.)

The problem of dealing with colored samples is also in reasonably good shape [3]. Here progress has been made by going to exciting lines other than Hg 4358 A. Some exciting lines which have been used and others which have been proposed are listed below.

| Available | | Proposed | |
|---|---|---|---|
| Hg | 5461 A | Rb | 7800 A |
| | 5770 A | | 7948 A |
| | 5790 A | Cs | 8521 A |
| He | 5876 A | | 8944 A |
| | 6678 A | Li | 6104 A |
| | 7065 A | | 6708 A |
| | 7281 A | | |
| Cd | 6439 A | | |
| Na | 5890 A | | |
| | 5896 A | | |
| K | 7665 A | | |
| | 7700 A | | |

In going to long-wavelength exciting lines, a number of new problems are introduced. In the first place, the intensity of Raman scattering depends on the inverse fourth power of the exciting wavelength. For equal source brightness, for example, scattering from He 5877 A will be only 0.3 as intense as that from Hg 4358 A. To compound this difficulty, photographic plates for the red region are perhaps 0.1 as sensitive as those for the blue region. At first blush, then, it would seem that exposure times might run 30 times as long in the red as in the blue region of the spectrum. Fortunately, there are factors which more than compensate for this. The foremost of these is the resonance Raman effect: as an exciting line moves into an absorption line of the sample from the long-wavelength side, the intensity of Raman scattering increases by a very large factor. In other words, it is disadvantageous to get rid of our color problem entirely. There is clearly some optimum wavelength to use. Our choice of exciting wavelengths is somewhat limited, but at a given wavelength there is an optimum solution concentration to give a maximum yield of Raman radiation. For strong absorbers, a solution may actually yield a stronger Raman spectrum than the pure liquid. That such an optimum concentration exists is difficult to show for a conical sample tube, but is shown with relative ease for the back-illuminated disc. From Fig. 2 we note that $\epsilon$ is proportional to the concentration and that the brightness of the sample, $B$, is also proportional to the concentration. The flux of Raman radiation is readily shown to be maximized when the concentration equals $1/(\epsilon a)$. The problem of optimizing both exciting wavelength and concentration is complicated and is generally tackled by hit-or-miss methods.

It is not amiss to say something about sources and spectrographs for colored exciting lines. Samples which can be excited with the Hg 5461 A line can be studied with a conventional spectrograph fitted with an $NdCl_3$ – Wratten-filter combination. The dispersion of prism spectrographs becomes increasingly poor in the yellow and red region. Raman work in this region is best done with a suitably blazed grating spectrograph. Although sodium, potassium, and cadmium lamps [4-6] have been used as sources, the most popular at present is the helium lamp [3,7]. This lamp has four useful lines at 5876, 6678, 7065, and 7281 A. The lamps are easily powered by a microwave generator or a high-voltage transformer. The big disadvantages of helium lamps are the difficulty of freeing the helium gas from traces of neon and the inefficient conversion of electrical to visible radiant energy. From this latter point of view, alkali metals are particularly advantageous; about 90% of the radiant output is in the resonance lines, and the ionization energies are low. Lithium looks particularly promising as a Raman source material, if a lamp can be developed. The new lasers being developed may also prove useful as Raman sources [8].

Some experiments conducted at the Cyanamid Laboratories indicate that fluorescence is radically reduced when the He 5876 A line is used as an exciting line, so that the solution to this problem may also lie in the direction of longer-wavelength excitation. (A Wratten #22 filter nicely removes wavelengths below 5500 A.)

Figure 8 shows a spectrum of aqueous chloroplatinic acid taken with the yellow mercury lines. While the exposure is long (12 hr), the plate is of excellent quality.

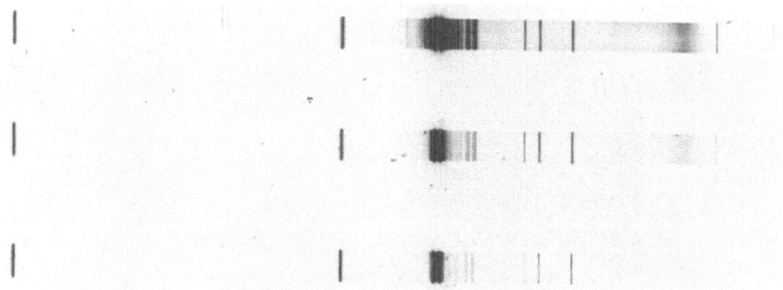

Fig. 8. The Raman spectrum of a $H_2PtCl_6$ aqueous solution. Hg 5770–5790 A excitation, 10-hr exposure (top), polarized light, $E \parallel$ axis (middle), $E \perp$ axis (bottom) (courtesy of Dr. R. F. Stamm).

The question naturally arises as to the possibility of obtaining spectra of turbid colored samples. This has, in fact, been done by Stammreich and Sala [9], who obtained spectra of turbid $[Mo(CN)_8]^{4-}$ solutions by using He 5876 A and He 6678 A as exciting lines and by using neodymium nitrate for the former and erbium perchlorate for the latter as bandstop filters. There seems to be no reason why bandstop interference filter units could not be constructed for long-wavelength lines.

In summing up, it seems clear that it is now feasible to construct Raman sources which get around most of the sample problem if photographic recording is used. There seems to be no reason why samples of 0.5 to 1.0 g of material could not be examined routinely in industrial or academic laboratories. Unfortunately, there still exists no photodetector for the green, yellow, or red comparable in sensitivity to the IP21 photomultiplier for the blue. When such a photo-detector is developed, we may expect to see, if not Raman Infracords, then Raman Model 21s or IR-4s become commonplace.

## REFERENCES

1.  M. C. Tobin, *J. Opt. Soc. Am.* **49**, 850, 1959.
2.  J. Brandmüller, *Z. angew. Physik* **5**, 95, 1953.
3.  H. Stammreich, *Spectrochimica Acta* **8**, 41, 1956.
4.  F. T. King and E. R. Lippincott, *J. Am. Chem. Soc.* **78**, 4192, 1956.
5.  F. A. Miller, G. Carlson, and W. White, *Spectrochimica Acta* **15**, 709, 1959.
6.  P. Krishnamurti, *Indian J. Physics* **5**, 587, 1930.
7.  N. Ham and A. Walsh, *Spectrochimica Acta* **12**, 88, 1958.
8.  S. Porto and D. Wood, *J. Opt. Soc. Am.* **52**, 251, 1962.
9.  H. Stammreich and O. Sala, *Zeit. für Elektrochemie,* **64**, 741, 1960.

# Advances in Ionization Detectors—
# The Electron-Affinity Detector

## S. J. Clark

Jarrell-Ash Company
Newtonville, Massachusetts

This paper reviews some of the recent work on the electron-affinity gas-chromatographic ionization detector. This apparatus, together with certain special applications, will be described briefly.

The measurement of the concentration of gas eluted from a chromatographic column may be achieved by one of several ionization reactions. Of these, one of the more important is ionization in the presence of argon carrier gas, this reaction providing the basis for the argon-ionization detector of Lovelock [1-3]. Subsequent work by the same author [4,5] has made it apparent that, by suitable changes in geometry and operating conditions, a series of detectors can be produced, each having well-defined characteristics and each lending itself to particular analytical applications. This paper is a preliminary report on some characteristics of one of these detectors, a commercial version of the electron-affinity detector [6-8].

Practical ionization detectors are usually operated at atmospheric pressure where the main process responsible for loss of ions is the recombination of negative and positive ions to form neutral molecules. With most detectors, ion recombination phenomena are undesirable since they set a limit to or interfere with the faithful response of the detector. The electron-affinity detector, however, is designed to exploit ion recombination for the detection of components having an affinity for free electrons.

## PRINCIPLES OF OPERATION

Stable negative ions may be formed by the capture of free electrons by neutral molecules according to the following reactions

$$xy + e^- \leftrightarrow xy^- + \text{energy}$$

$$xy + e^- \leftrightarrow x + y^- \pm \text{energy}$$

If the ionization chamber contains only the source of ionizing radiation and a carrier gas having no affinity for electrons, the charge carriers in the ionized gas will be positive molecular ions and electrons. The extremely high mobility of free electrons makes ion recombination a very unlikely process and, thus, the application of a small potential to the chamber suffices to collect all the charge carriers formed in the ionization reaction.

If, now, a gas having an affinity for free electrons is mixed with the carrier gas and introduced into the chamber, formation of negative molecular ions will occur through electron capture. The mobilities of negative molecular ions are very much less than those of electrons and, consequently, the probability of ion recombination is very much enhanced in the presence of these ions. Thus, the introduction of an electron-capturing substance into the ionization chamber will result in a decrease in the ion concentration in the gas and a decrease in the current flowing in the chamber.

As the potential across the chamber is increased, the residence time of the ions in the chamber decreases; thus, the probability for both electron capture and ion recombination decreases. At sufficiently high potentials, both electron capture and ion recombination virtually cease to occur and the saturation current again flows in the chamber.

In a detector having plane-parallel geometry, the fractional ion loss is given by the expression [9]

$$f = xNd^4/\lambda^+\lambda^- V^2$$

where $f$ is the fractional ion loss, $x$ is the recombination coefficient (related to the electron affinity of the molecule), $N$ is the rate of ion production (related to concentration of gas), $d$ is the distance between the electrodes, $\lambda^+$, $\lambda^-$ are the mobilities of positive and negative ions, respectively, and $V$ is the applied potential.

From the above equation it can be seen that the applied potential required to collect a given proportion of the ions formed is dependent on the recombination coefficient and, thus, upon the electron affinity or absorptivity of the molecule. Thus, it becomes possible to distinguish between molecules having different electron affinities or, alternatively, to detect traces of strongly capturing compounds in the presence of large amounts of material having a low affinity for electrons.

It should be realized that the above explanation is considerably simplified and that the processes occurring are considerably more complex than has been indicated. However, under the relatively constant conditions obtaining in gas chromatography, reliable results can be achieved using relatively simple apparatus.

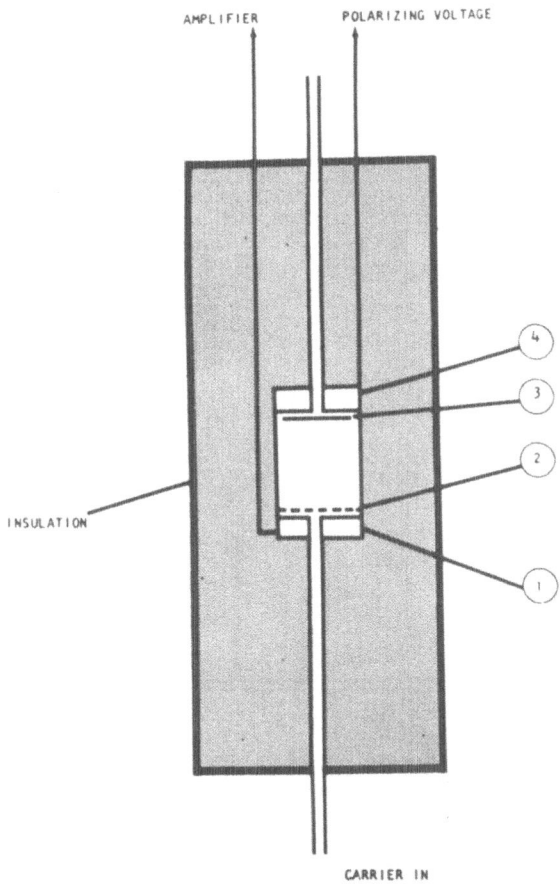

Fig. 1. Jarrell-Ash Electron Affinity Detector: (1) anode, (2) wire screen, (3) radioactive source, (4) cathode.

## APPARATUS AND METHODS

A Jarrell-Ash Electron-Affinity Detector (Catalog No. 26-755) was used throughout this work. The detector design follows that of Lovelock [5] and its construction is illustrated in Fig. 1. Carrier gas is admitted to the chamber through the anode (1), the wire screen (2) covering the entrance serving to diffuse the gas into the chamber and prevent turbulence. The gas stream leaves the chamber through the cathode (4), to which is attached the radioactive source (3), a stainless steel foil coated with titanium tritide. The activity of the foil is such as to provide a dc saturation current of 3 to $6 \cdot 10^{-9}$ amp. The polarizing voltage is supplied by a stabilized dc supply having an output continuously variable over the range 0–85 v (Catalog No. 26-770). The detector signal is amplified by an electrometer amplifier (Catalog No. 26-770) and recorded by a standard 10 mv potentiometric strip chart recorder.

A flow diagram of the chromatographic apparatus is shown in Fig. 2. For best performance, the gas flow rate through the detector should be in the range of 100–200 ml/min. If necessary, diluent gas is added to achieve flow rates in this range through the "T" located between column and detector. The detector temperature is maintained constant at 220°C by a thermostatted heating block (Catalog No. 26-750).

The results reported below were obtained using standard 4-ft, $^1/_4$-in.-o.d. $^3/_{16}$-in.-i.d. stainless steel columns. Column packings consisted of 80–100 mesh Chromosorb W coated with 5% by weight of

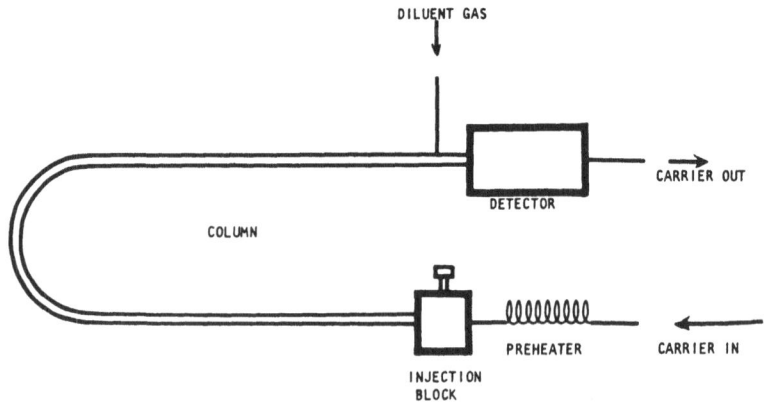

Fig. 2. Flow diagram of chromatographic apparatus.

the appropriate stationary phase, or of 140- to 200-mesh glass beads coated with approximately 0.2% of stationary phase.

## QUALITATIVE APPLICATIONS

It is evident from the relationship between ion loss and applied potential given above that the response to a particular compound will vary with applied potential. At the lowest potentials, ion loss through electron capture gives rise to a negative signal, the magnitude of the signal being dependent upon the electron absorptivity of the compound. As the potential is increased, the signal decreases until no change in current is observed. At still higher potentials, the detector begins to function as a normal, though inefficient, ionization detector and a positive signal is obtained. The range of voltage over which this transition occurs will depend upon the electron absorptivity of the molecule, which is itself a function of molecular type. In principle, then, a series of chromatograms recorded at different applied potentials serves to distinguish between molecular types. This information, together with a knowledge of retention times, is often sufficient to identify the components of a mixture. A chromatogram illustrative of this use of the detector is shown in Fig. 3.

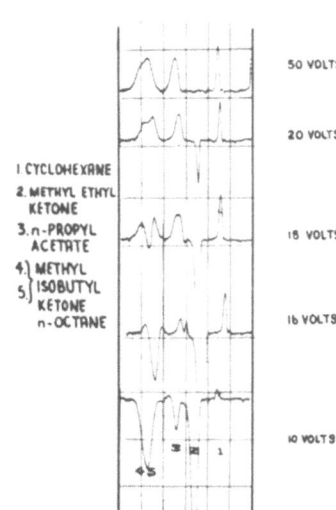

Fig. 3. Chromatogram.

This mode of operation has several disadvantages however. First, it is necessary to run several chromatograms to obtain sufficient information for the analysis. Second, the mean energy of electrons in a gas depends upon the nature of the gas. Thus, the electron energy will change when sample gas enters the detector and, moreover, the extent of the change will be dependent upon the concentration of the sample in the carrier gas. Hence, the response of a given compound at a particular voltage will be concentration-dependent. This dependence may cause uncertainty in the interpretation of the chromatogram.

More consistent results are obtained if the detector is operated at a fixed potential, just sufficient to produce saturation in the detector with pure carrier gas flowing. Under these conditions, the detector will operate near maximum sensitivity for strongly capturing compounds, although compounds having low electron absorptivities may well produce positive signals, particularly when argon is used as the carrier gas. This mode of operation does not allow identification of compound type from electron absorption characteristics unless the concentration of the individual components of a mixture is known. This information can be obtained by connecting a quantitative detector in parallel with the electron-affinity detector. The effluent from the chromatographic column is divided between the two detectors and simultaneous chromatograms are recorded. The differing response of the two detectors is then used to provide a measure of the electron absorptivities of the components of the mixture. Figure 4 illustrates the analysis of a solvent mixture using the electron-affinity detector in parallel with an argon diode. The argon diode tends to respond anomalously to very strongly capturing compounds; a better detector for the purpose would be the photoionization detector of Lovelock [10] in which ion recombination does not occur or, possibly, the argon diode operated at reduced pressure.

Work is in progress to determine electron absorptivities of a wide range of compounds; this information will facilitate identification with the system just described. In general, hydrocarbons, alcohols, esters, ethers, and amines have low absorptivities, whereas ketones, halogenated compounds, quinones, acid anhydrides, and nitro compounds absorb strongly.

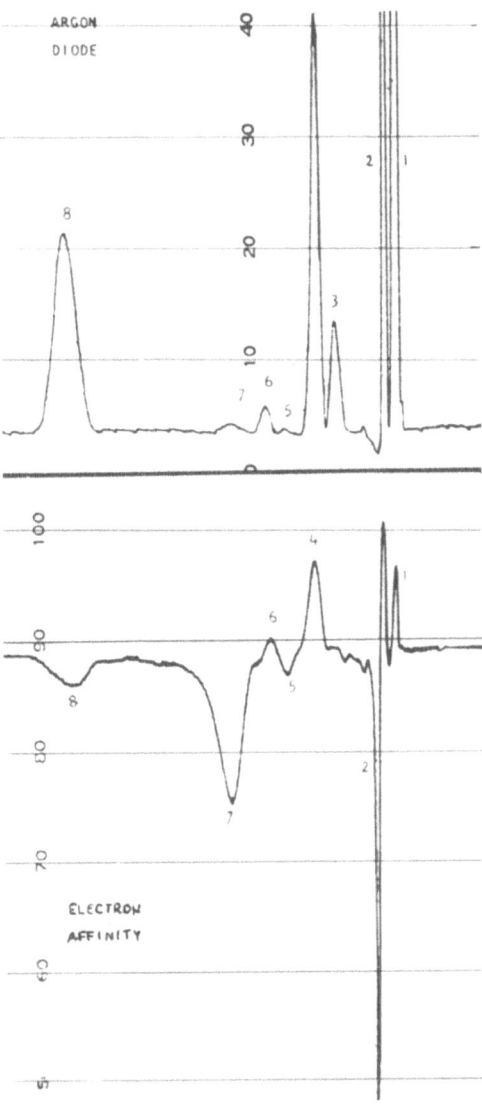

Fig. 4. Analysis of a solvent mixture: (1) hydrocarbon, (2) ethyl acetate, (3) isobutanol, (4) hydrocarbon, (5) methyl ethyl ketone, (6) hydrocarbon, (7) methyl isobutyl ketone, (8) butyl cellosolve. Column: diethylene glycol succinate; 80°C; argon, 100 ml/min.

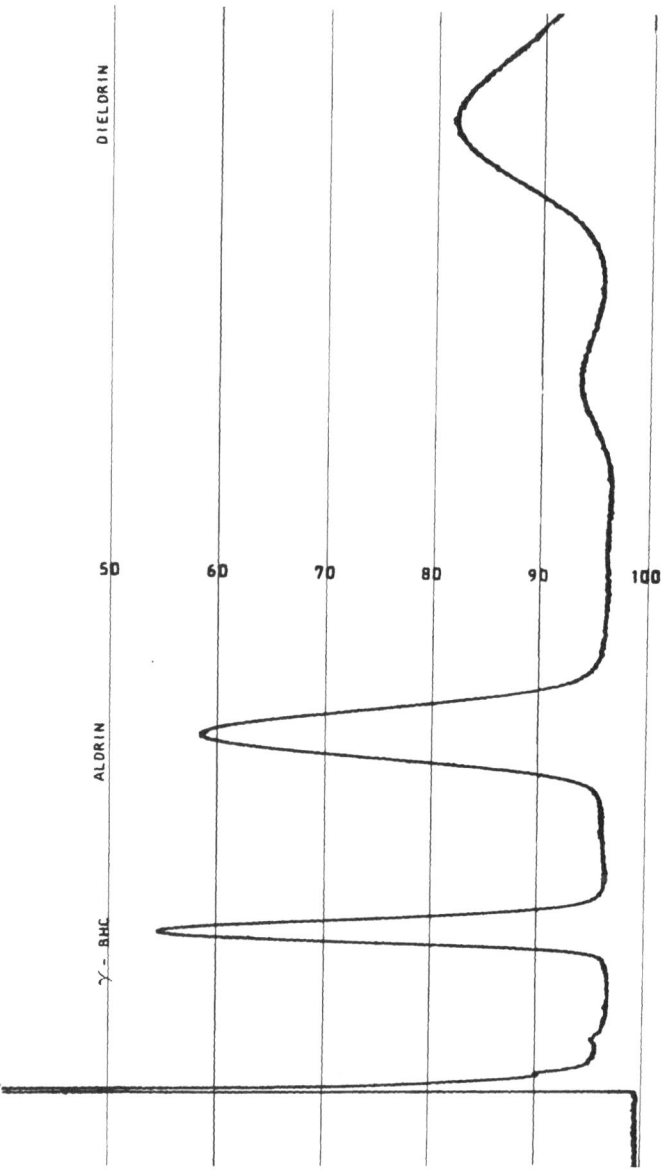

Fig. 5. Chromatogram of mixture of chlorinated pesticides. Column: silicone grease; 165°C; nitrogen, 80 ml/min. Components, $10^{-10}$ g each.

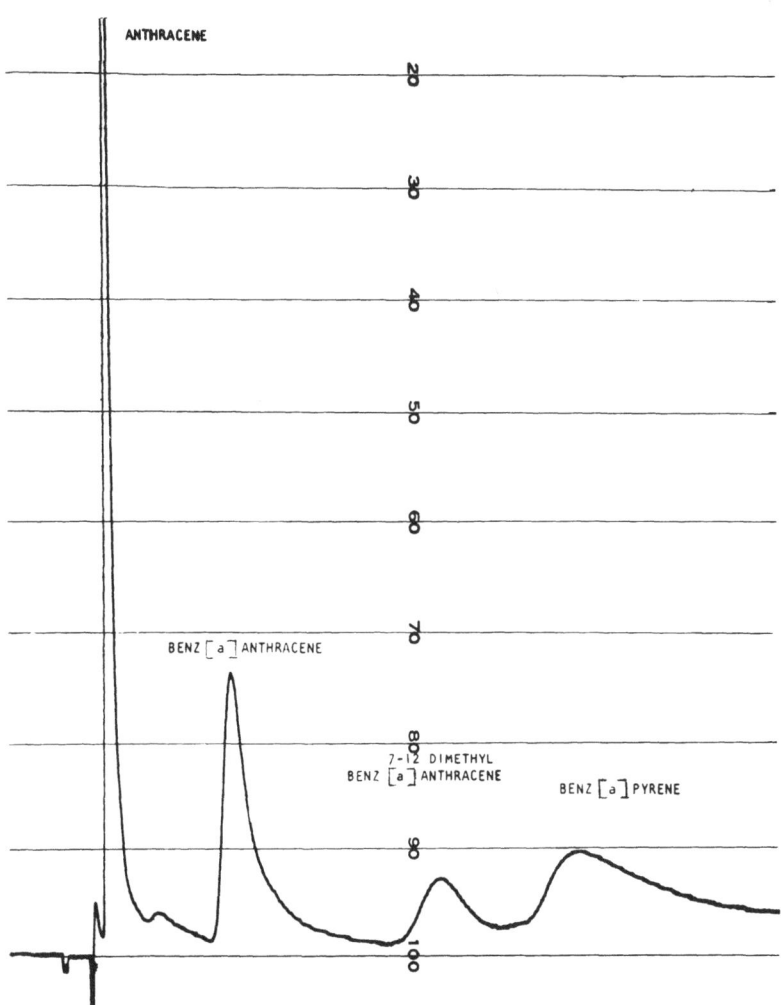

Fig. 6. Chromatogram of mixture of polycyclic hydrocarbons. Column: silicone grease on glass beads; 200°C; nitrogen, 130 ml/min. Components, approximately $10^{-9}$ g each.

## QUANTITATIVE APPLICATIONS

Little information is yet available concerning the quantitative use of the detector, but studies so far indicate that for certain important groups of compounds it will prove to be an extremely valuable tool of unequalled sensitivity. Since the detector is sensitive, it becomes possible to determine traces of strongly capturing compounds in the presence of large amounts of weakly absorbing compounds. Figures 5 and 6 show chromatograms of mixtures of chlorinated pesticides and of polycyclic hydrocarbons. The extreme sensitivity of the analyses illustrates the potential value of the technique for the determination of traces of toxic materials in the analysis of foodstuffs or of air pollutants. The present limit of detection for the chlorinated pesticides and the polycyclic hydrocarbons is approximately $10^{-12}$ and $10^{-10}$ g, respectively, with some variation depending upon the individual compound. It seems likely that these figures can be improved by one or two orders of magnitude by appropriate changes in the experimental conditions.

The response of the detector operated under dc conditions is linear over 10–20% of its working range. Figure 7 shows a calibration curve for the chlorinated pesticide, methoxychlor.

Lovelock [8] has recently introduced a modification of the system in which the charge carriers are collected by applying pulses of about 1 $\mu$sec in width at intervals of 20–50 $\mu$sec. Under these conditions, the response of the detector follows a Beer's Law type of relationship

$$I = I_0 e^{-ack}$$

where $I_0$ is the detector saturation current, $a$ is the electron absorptivity of the compound, $c$ is the concentration, and $k$ is a constant depending upon detector geometry and operating conditions.

This relationship emphasizes the fact that electron absorption may be regarded as an analog of photon absorption and that the technique is, in fact, a form of spectroscopy. The electron-capture reaction is a resonance process, for there are no product electrons to carry away excess energy. This is true even when capture brings about dissociation, since the range of atom and ion kinetic energies is finite. Thus, if monoenergetic electrons can be produced in a suitable ionization chamber, the possibility exists for resolution of

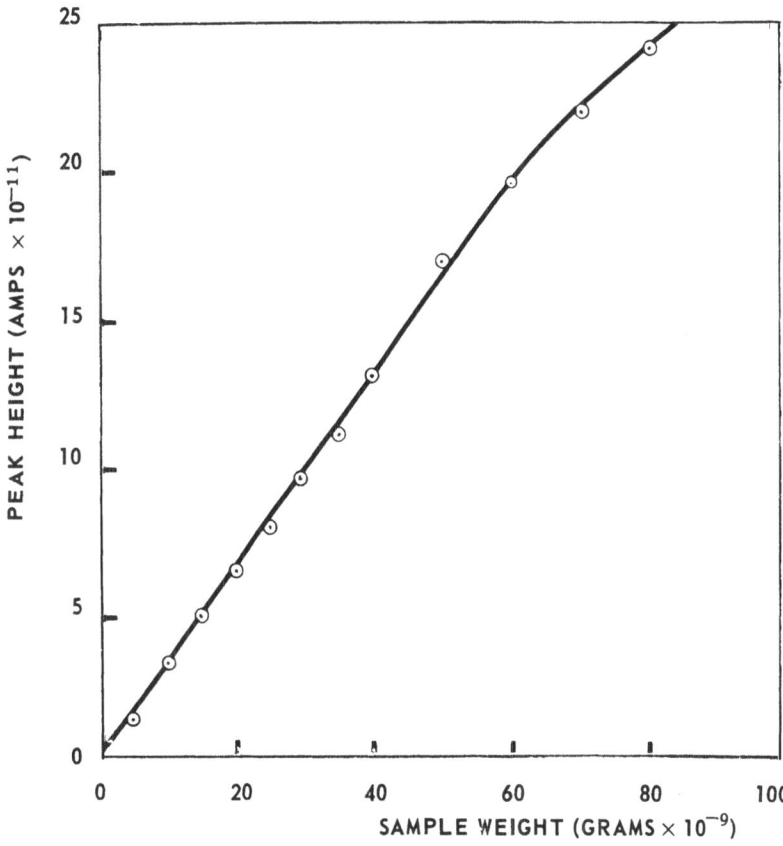

Fig. 7. Calibration curve for methoxychlor.

electron absorption spectra comparable with that achieved for other forms of spectroscopy.

## REFERENCES

1.  J. E. Lovelock, *J. Chromatog.* 1, 25, 1958.
2.  J. E. Lovelock, A. T. James, and E. A. Piper, *Ann. N. Y. Acad. Sci.* **72**, 720, 1959.
3.  J. E. Lovelock, *Nature* **181**, 1460, 1958.
4.  J. E. Lovelock, *Gas Chromatography*, R. P. W. Scott, ed., Butterworths, London, 1960, p. 16.
5.  J. E. Lovelock, *Analytical Chemistry* **33**, 162, 1961.
6.  J. E. Lovelock and S. R. Lipsky, *J. Am. Chem. Soc.* **82**, 431, 1960.
7.  J. E. Lovelock, *Nature*, **189**, 729, 1961.
8.  J. E. Lovelock and N. L. Gregory, *Third International Symposium on Gas Chromatography*, E. Lansing, Michigan, June 1961.
9.  J. Sharpe, *Nuclear Radiation Detectors*, Methuens, London, 1955, p. 130.
10. J. E. Lovelock, *Nature*, **188**, 401, 1960.

# Techniques Utilized in Studying the Structures of Inorganic and Coordination Compounds

Herman A. Szymanski

Canisius College
Buffalo, New York

The determination of the infrared spectra of inorganic and addition compounds sometimes requires special techniques. In addition, the combination of several spectrophotometric methods is sometimes required in order to completely determine structures.

We shall report some of the techniques used in our work as well as the results obtained. Included will be a discussion of rapid sublimation techniques, a comparison of results obtained with the several standard techniques, and a description of the materials used, such as $D_2O$, to reduce scattered light.

A discussion of peak shifts which occur in some inorganic compounds when their environment is changed will also be made.

A brief discussion on the correlation of infrared data with nuclear magnetic resonance spectroscopy will also be made.

I should like to report a number of techniques my students and I have utilized in using spectroscopy to determine the structures of inorganic and coordination compounds.

## RAPID SUBLIMATION TECHNIQUES
## FOR REDUCING PARTICLE SIZE OF SAMPLES

It is necessary to have the particle size of a sample below the wavelength of light used to measure its absorption spectra. For infrared spectra, this particle size must, therefore, be below $3\mu$; otherwise a good deal of scattered light can occur, resulting in poor spectra. Generally, inorganic materials are hard and cannot be easily reduced to a size for which the effect of scattered light is small. We recently reported a technique in which the material is sublimed very quickly from a platinum dish onto the cold salt plate [¹]. This was accomplished by using an induction furnace which caused sublimation to take place in a few seconds. A very fine powder was deposited onto the salt plate and gave spectra which showed very little scattered light. This technique can be extended to coordination compounds containing organic linkages and even to some organic compounds. Using an evacuated infrared gas cell, so no combustion could take place, we have sublimed complexes such as those formed from tertiary amines with arsenic and antimony halides onto the end windows of the cell. No destruction of the organic part of the complex takes place during the sublimation. For one complex, dimethylaniline–antimony trichloride, we were able to show that this technique was better than conventional sublimation techniques. In these conventional techniques the infrared spectra revealed decomposition products which did not appear in our spectra.

Finally, while we have not attempted to study sublimation of organic compounds extensively, we did sublime several with good success. One of these, caprolactam, showed a very sharp infrared spectrum compared to that measured for it in mull oil.

## OTHER TECHNIQUES UTILIZED
## IN REDUCING SCATTERED LIGHT IN SAMPLES

The problem of scattered light has occurred in much of our work with inorganic and coordination compounds. In some work we have done on zeolites (trade name of sieves), we were interested in physically as well as chemically sorbed materials. While infrared spectra can easily distinguish between the two types, preparing the sample to obtain this information from its spectra is usually quite difficult. I would like to report two techniques used in solving this problem.

The large attraction zeolites have for water makes it difficult to remove the last traces of water from them. In addition, as the last traces of water are removed from a zeolite, the scattered light the sample shows increases a great deal, making it difficult to obtain a good infrared spectrum. One method we used to prevent zeolites from picking up water while their spectra were being measured, and to reduce the scattered light, was to quench the heated zeolite in mull oils. The oil prevented further water from being sorbed and was also used as the dispersing agent in which the spectrum was run.

Another technique utilized to reduce scattered light was replacement of physically sorbed water by deuterium oxide. Any chemically sorbed water would not be replaced by this oxide; thus we were able to show the presence of surface OH groups on the zeolite that could not be seen in the spectra of the dry powder, although they were present there.

We should like to suggest the use of a special spatula covered with 400 mesh platinum to grind very hard particles. We have been successful in using it to obtain very fine particle size with zeolites which cannot be easily ground in a conventional mortar.

## TECHNIQUES UTILIZED IN STUDYING THE REACTION OF TERTIARY AMINES WITH ARSENIC TRICHLORIDE

Conventional sampling techniques, such as mull oils and potassium bromide pellets, are difficult to use for very hygroscopic materials. Having used them to study the reaction between tertiary amines and arsenic trichloride, we decided that other techniques were required to completely identify all the phases of this reaction. The work was then done in vacuum lines so that no contact with moisture was possible. A specially designed infrared cell was attached to the vacuum line and the samples transferred to this cell. A diagram of this cell is shown in Fig. 1. Since we wished to use a liquid medium, it was decided that the acceptor molecule, arsenic trichloride, could be used as the solvent. This gave us a second advantage since the same liquid could be used as the solvent for nuclear magnetic resonance (nmr) studies. Thus, high-resolution nmr could be combined with infrared and the results compared. Sampes for nmr study were taken from the sample vacuum line at the same time as those for infrared analysis. Thus, we were measuring the same sample by nmr and infrared. We did obtain an interesting result in this comparison. The infrared data indicated that about 30% of amine hydrochloride was present in the solution but the nmr spectrum did not. Comparing the nmr spectrum to that of amine hydrochloride showed no lines of similar character. The only reasonable explanation for this is that the hydrochloride is present but is in rapid exchange with another species in solution. In this case, then, the hydrochloride lines would not be present in the nmr spectra but lines representing

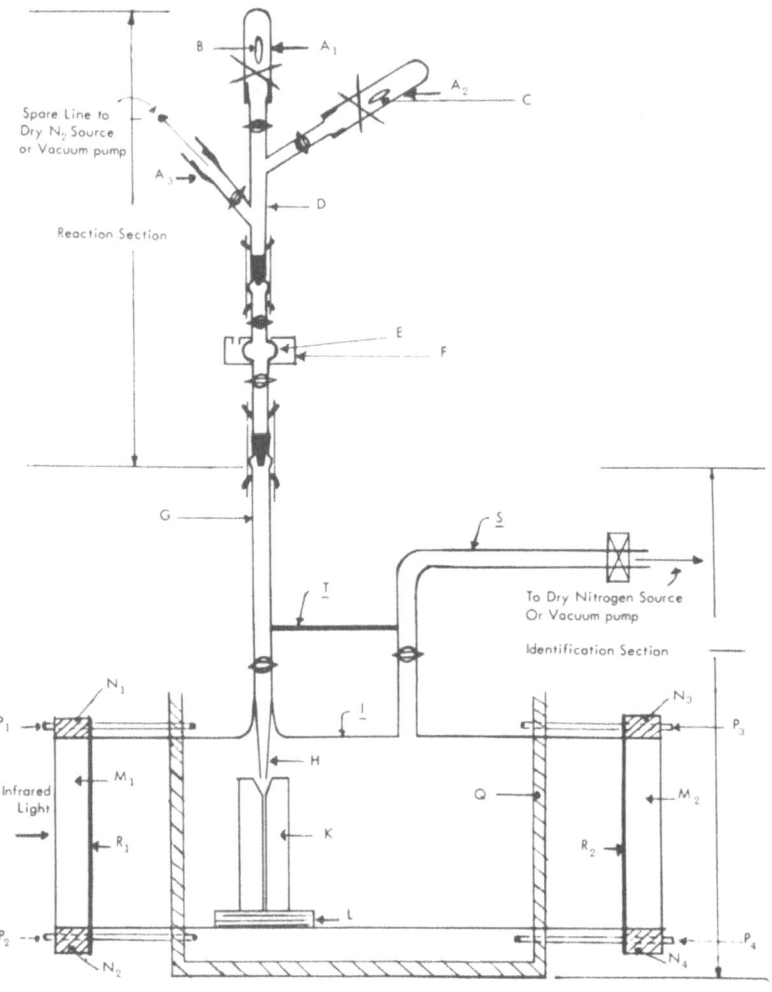

Fig. 1. Infrared identification apparatus. $A_1$, $A_2$, polyethylene tubing closed at one end; $A_3$, spare line to dry $N_2$ source or vacuum pump; $B$, ampoule of arsenic trichloride; $C$, representative of an ampoule of trimethylamine; $D$, glass tree complete with teflon stopcocks (3), 10/30 male ground-glass joint with glass tips; $E$, reaction flask (8 to 10 ml capacity) complete with stopcocks, a 10/30 female ground-glass joint to be fitted to its counterpart on $D$; $F$, cold box surrounding reaction flask; $G$, glass tube about 1 ft long, complete with 10/30 female ground-glass joint at its upper end to fit its counterpart on $E$ and a stopcock near its lower end; $H$, glass spout connected to $G$; $I$, glass cell about $1\frac{1}{2}$ inches in diameter;

an average of the hydrochloride and the species in equilibrium with it would be seen. This phenomenon does occur quite frequently.

Our infrared spectral data still do not give us all the necessary information concerning the reaction of amines with arsenic trichloride. We were able, however, to establish that at least three products were formed in this reaction.

This was shown by varying the method of preparation and noting the relative changes in peak intensity which occurred. By this technique, it was possible to show that one component gave bands at 1090 and 1000 cm$^{-1}$ while a second had bands at 1040 and 990 cm$^{-1}$. The third component gave bands at 1050 and 975 cm$^{-1}$. They all appear to be somewhat similar in structure since their infrared bands are quite similar. It was possible to obtain two of the three in pure form. Their complete infrared spectra showed that bands in other positions were quite similar and only those bands listed above could be used to distinguish each component.

## REFERENCES

1. H. A. Szymanski and P. Peller, *Applied Spectroscopy* 14, 4, 107, 1960.

K, small sodium chloride sandwich cell, either 0.1 mm or 0.5 mm thick; L, small plastic platform; $M_1$, $M_2$, sodium chloride salt plates fitted to the ends of the glass cell; $N_1N_2$, steel holder plates (holder for the $N_3$, $N_4$ sodium chloride salt plates); $P_1$, $P_2$, $P_3$, $P_4$, bolts; Q, stainless steel box reservoir fitted around the glass cell; $R_1$, $R_2$, silicone rubber gasketing; S, glass tube fitted with stopcocks—connection to either dry nitrogen source or vacuum pump; T, glass connection between glass tubes.

# The Visible and Near-Infrared Absorption Spectra of Some Trivalent Actinide and Lanthanide Elements in DClO₄ and in Molten Nitrate Salts*

W. T. Carnall and P. R. Fields

Argonne National Laboratory
Argonne, Illinois

Molten salts constitute an important class of solvents for the study of the chemistry of the actinide and lanthanide elements, particularly the intensely radioactive transuranium elements where the stability of certain of these systems to radiation decomposition is important. One tool useful in this type of study is the observation of absorption spectra. In the present investigation we report observations of absorption spectra of trivalent actinide and lanthanide ions in $LiNO_3$–$KNO_3$ eutectic at 150°C in the region between 0.36 and 2.5 $\mu$. Transitions to low-lying excited states have been observed in crystals of both actinide and lanthanide elements in the near-infrared region (1.3–2.5$\mu$) and thus would be predicted to be observable in solution. However, the existence of intense solvent absorption bands in aqueous solution above 1.3 $\mu$ constitutes a severe limitation on the use of that medium. In the nitrate eutectic there are essentially no interfering solvent absorption bands between 1.3 and 2.5 $\mu$. It is also possible to observe portions of this region in deuterated solvents, and thus we are able to compare over an extended range the effects of different ligand fields on 4f and 5f electrons. The bands observed in the near-infrared region should prove to be of analytical importance since the extinction coefficients appear to be comparable to many of the more intense bands in the visible region.

The absorption spectra of the trivalent lanthanide elements have been of interest to both chemists and spectroscopists for many years. Chemists have manifested particular interest in the effects of different solvent media on the relatively sharp absorption bands seen

*Based on work performed under the auspices of the U.S. Atomic Energy Commission.

in the 0.2 to 1.35 $\mu$ region. Such studies are designed to characterize the relative complexing ability of various anionic species. The analytical importance of these absorption bands has also received considerable emphasis. Spectroscopists have, on the other hand, required well-characterized crystal lattices as matrices for the lanthanides, which when studied at low temperatures make possible the location and assignment of the many lines arising from the intra-4$f$ electron transitions. The complexity of the systems involved is attested by the continuing appearance in the literature of refinements and extensitons of earlier results. The production of synthetic transuranium elements has made possible similar studies on 5$f$ electron systems.

Molten salts provide an ionic, quasi-lattice matrix within which the concentration of solute species can be varied over a wide range. As media for the observation of the absorption spectra of the trivalent lantahanide and actinide elements, they are closely related to the solid state. Many molten salt systems are stable to radiation decomposition; in addition, chemical reactions can be conducted in them in such a manner that adaptations of commonly used separation procedures are possible. These properties make molten salts particularly attractive media for the study of the highly radioactive actinide elements.

## EXPERIMENTAL

The spectral measurements were made using a Cary Recording Spectrophotometer Model 14. The aqueous solutions were run at room temperature, $23 \pm 2°C$. For the molten salt samples, a furnace essentially identical to that designed by Young and White was used [1]. The temperature could be maintained at $150 \pm 3°C$ simply by balancing the energy input against the heat loss; this obviated the necessity for a control unit.

The molten salt matrix used was a eutectic mixture of $LiNO_3$ and $KNO_3$. The eutectic contained $ca.$ 43 mol% $LiNO_3$, giving a melting point of $ca.$ 132°C. Reagent grade chemicals were used without further purification. Proper proportions of dried salt were fused at $ca.$ 220°C and treated by bubbling dry $N_2$ through the melt for several hours to drive off residual $H_2O$. The system was next evacuated for 30 min, followed by filtration of the molten salt through a fine porosity pyrex filter; then the salt was molded into long sticks of 0.8 cm diameter. The latter, when broken into short lengths, were convenient for weighing. The supply was kept in a dry box. The technique of preparing $DClO_4$ has been described previously [2].

The lanthanide elements used were obtained commercially in the form of oxides, and contained less than 1% impurities as supplied. Samples of the dried oxides were weighed out and dissolved in the appropriate acid, $DClO_4$ or $HNO_3$. The actinide elements were taken from stocks at Argonne National Laboratory. Their purification and method of assay have been described previously [3]. Total impurities in each actinide sample were less than 1%, and consisted of noninterfering elements such as Al and Na.

The samples for the nitrate melt were dissolved in $HNO_3$ and evaporated to dryness in the spectrophotometer cell before adding the desired weight of $LiNO_3 - KNO_3$ eutectic. Cells for this purpose were made of precision-bore square Pyrex tubing. Each cell, ca. 3 ml in volume with a 1-cm path length, was sealed onto a length of 12-mm Pyrex tubing with a ground-glass joint at the other end. The overall length of the unit was ca. 30 cm. The long stem facilitated handling of the cells.

The wavelength limits within which Pyrex glass transmits essentially 100% of the incident light are very nearly the same limits as imposed by the nitrate eutectic. A blank run of the Pyrex cells with $CCl_4$ as the solvent revealed appreciable absorption below 0.3 $\mu$, but only slightly increasing absorption at 2.6 $\mu$, which is the long-wavelength limit of the instrument used. Florence and coworkers [4] indicate that the infrared cutoff of Pyrex glass, depending upon its pretreatment, is in the vicinity of 2.6 $\mu$.

Standard 1-cm-path-length silica cells equipped with a glass stopper were used for the deuterated solutions. Cells containing radioactive solutions were placed in a closed secondary container which took the place of the usual cell holder [3].

Extinction coefficients, $\epsilon = OD/C \times L$, where $OD$ is observed optical density, $C$ is concentration of the absorbing species in moles per liter, and $L$ is the path length of the cell in centimeters, were calculated for the molten $LiNO_3 - KNO_3$ using 1.948 g/ml as the density at 150°C, and assuming a similar density for the lanthanide or actinide nitrate dissolved. Since the solutions were dilute, the error involved in this assumption is probably less than 3–4%. In each case, a blank was run against air using a matched cell containing the same solvent as the sample cell. This was followed by running the sample

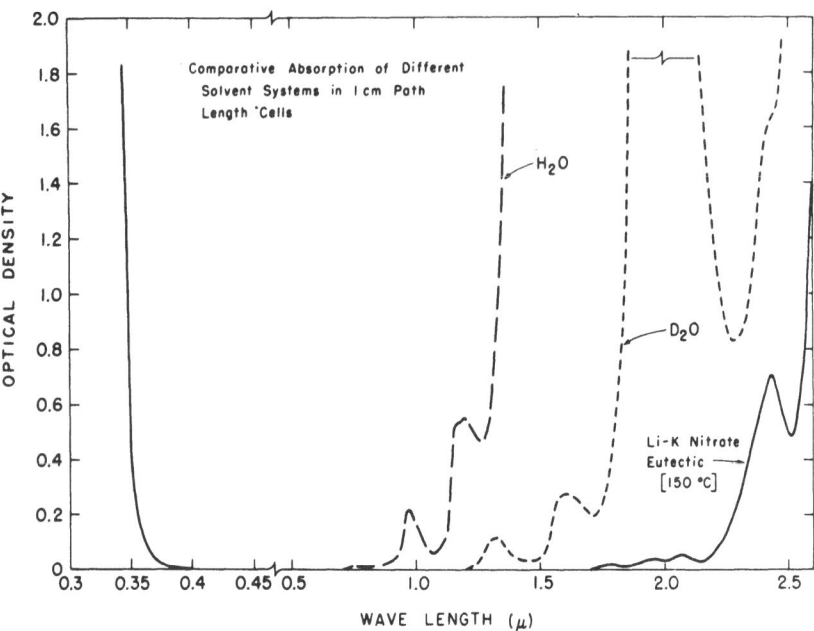

Fig. 1. Comparison of the visible and near-infrared absorption bands of $H_2O$, $D_2O$, and $LiNO_3$–$KNO_3$ eutectic.

cell *vs* air in the same spectral region. For the aqueous samples, runs of sample *vs* pure solvent in a matched cell were also made.

## DISCUSSION

Practically all of the investigations of the solution absorption spectra of trivalent lanthanide and actinide elements have been limited to the region between 0.2 and 1.35 $\mu$. As seen in Fig. 1, the near-infrared limit is imposed by the existence of strong $H_2O$ absorption bands. This limit is significantly extended by recourse to deuterated solvents, but again strong solvent absorption bands exist near 2.0 $\mu$. Except for a partial "window" from 2.12 to 2.37 $\mu$, observations in a 1-cm-path-length cell in $D_2O$ beyond *ca.* 1.8 $\mu$ must be made over an appreciable background absorption. In contrast, the molten $LiNO_3$ – $KNO_3$ eutectic at 150°C exhibits essentially no absorption in the 0.36–2.5 $\mu$ range, except for the rather broad band near 2.4 $\mu$. The lower wavelength limit is imposed by a $NO_3^-$ absorption band with center near 0.290 $\mu$ [5]. At wavelengths longer than 2.6 $\mu$, the overtones of the various vibration frequencies of $NO_3^-$ exhibit more intense absorption [4,6,7]. Below 2.7 $\mu$ all the $NO_3^-$ absorption bands are due to overtones of hydrogenic stretching vibrations or a combination of hydrogenic stretching and other modes of vibration in the molecule. Thus, in using 1-cm-path-length cells, the nitrate eutectic is a useful spectrophotometric solvent over the entire visible and near-infrared range up to 2.6 $\mu$. The relatively low melting point (132°C) can be attained with very simple furnace arrangements and, thus, most spectrophotometers should be easily adaptable to this type of solvent system.

Taken together, results in $DClO_4$ and in the nitrate eutectic make possible a comparison of the spectra of the hydrated free metal ions with that obtained under anhydrous conditions with nitrate ions responsible for the ligand field experienced by the metal ion. The extension of measurements into the near-infrared region makes possible the comparison of intensities of the bands found with those in the visible region. While these bands have previously been observed in crystal studies, a direct comparison in intensity has not been made. For the purpose of this communication, the discussion of lanthanide spectra will be limited to those light lanthanides possessing 4$f$ electrons when in the trivalent state in the nitrate melt. These elements are, then, the analogues of the actinide elements most readily available at the present time.

## LANTHANIDE SPECTRA

Within the limitation mentioned above, the species studied were $Pr^{3+}$, $Nd^{3+}$, $Sm^{3+}$, and $Eu^{3+}$. Trivalent cerium possesses one 4$f$ electron; however, in the nitrate melt, $Ce^{3+}$ is unstable. It is readily oxidized to $Ce^{4+}$. Promethium, the only lanthanide not found in nature, was not available for the present study. However, as a result of its relatively short half-life and, thus, its high specific activity, the molten nitrate system would be an interesting matrix in which to view the absorption spectrum of this element. An investigation of this type will be carried out in the near future.

A recent study of the solution absorption spectrum of $Pm^{3+}$ in DCl failed to reveal the existence of any bands in the 0.85–1.8 $\mu$ region [8]. It was, however, shown, in agreement with earlier work [9], that many of the $Pm^{3+}$ absorption bands possess very low extinction coefficients. Thus, low intensity bands in the near-infrared might be difficult to distinguish from the background, particularly since the

high specific activity of the solute severely limits the concentrations
at which it is reasonable to carry out investigations. Recent work
on the luminescence spectrum of $PmCl_3$ indicated lines in the visible
region and at 0.742 and 0.830 $\mu$; however, no longer-wavelength lines
were reported [10].

Calculation of the multiplet levels of $Pm^{3+}$ based on pure $(LS)$
coupling, using a value for the Slater integral, $F_2 = 334.8$ cm$^{-1}$, a
spin-orbit coupling constant, $\zeta_{4f} = 1015$ cm$^{-1}$, and spin-orbit splitting factors as tabulated by Elliott, Judd, and Runciman [11], pre-
dicts near-infrared bands at approximately 0.9, 1.0, 1.1, 1.5, and 2.2
$\mu$. If these bands exist, it may be possible to locate them using rather
high concentrations of $Pm^{3+}$ in the nitrate melt.

Gadolinium has no visible or near-infrared absorption bands.
The first excited multiplet level over the $^8S$ ground state is $^6P_{7/2}$,
which lies in the ultraviolet near 0.3119 $\mu$ [12].

The spectra of $Pr^{3+}$, $Nd^{3+}$, $Sm^{3+}$, and $Eu^{3+}$ in molten
$LiNO_3$ – $KNO_3$ eutectic at 150°C, as compared to the bands in dilute
$DClO_4$ at 25°C, are shown in Figs. 2, 3, 4, and 5. The intensities of
the bands in $DClO_4$ relative to those in the nitrate melt in the figures
result from an arbitrary selection of concentrations and should not
be directly interpreted in terms of extinction. The results in Table I
are compared on the basis of extinction coefficients for some of the
major peaks of the absorbing species in the two solvent systems. The
general trend appears to be toward lower intensity bands in the
molten nitrate, although in terms of oscillator strength this may not
be the case since the half-width of the bands in the melt appears to
be greater. There are, of course, clear cases of an increase in intensity
of a given band in the melt as compared to that in $DClO_4$, as, for
example, the band at 0.465 $\mu$ in $Eu^{3+}$, and that at 0.582 $\mu$ in $Nd^{3+}$.
The different distribution of the population within the various Stark
levels of the ground state occasioned by comparing spectra obtained
at 150°C with that obtained at 25°C, and, thus, the different transition probabilities obtaining would lead one to expect a shift in the
intensity pattern between the two temperatures as well as in the
center of gravity of the bands observed.

The gross similarities in the spectra in the two different solvents
are in accord with the results of Gruen and coworkers [13] for the
transition elements $Co^{2+}$, $Ni^{2+}$, and $Cu^{2+}$ in the $LiNO_3$ – $KNO_3$
eutectic as compared to spectra exhibited in dilute acid. In the aqueous solution, the ligand field experienced by the central cation is de-

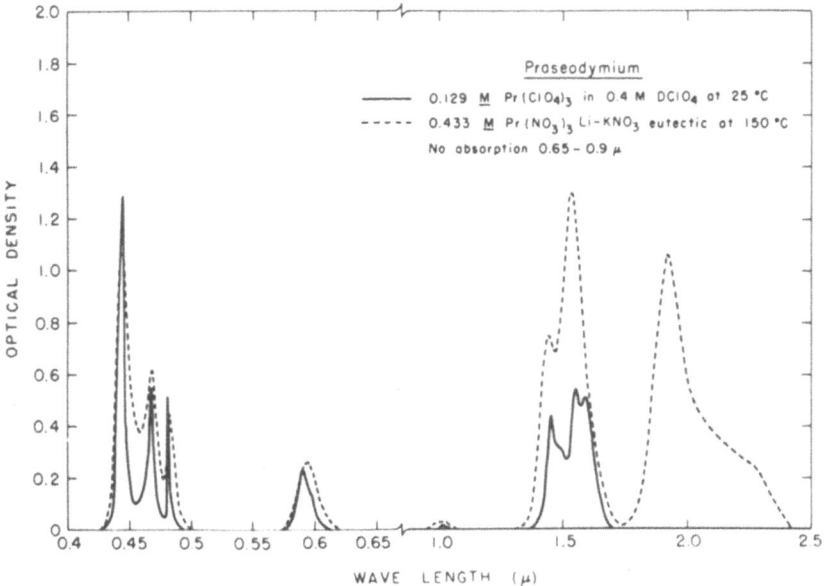

Fig. 2. Absorption spectrum of $Pr^{3+}$ in dilute $DClO_4$ at 25°C, and in molten $LiNO_3 - KNO_3$ eutectic at 150°C.

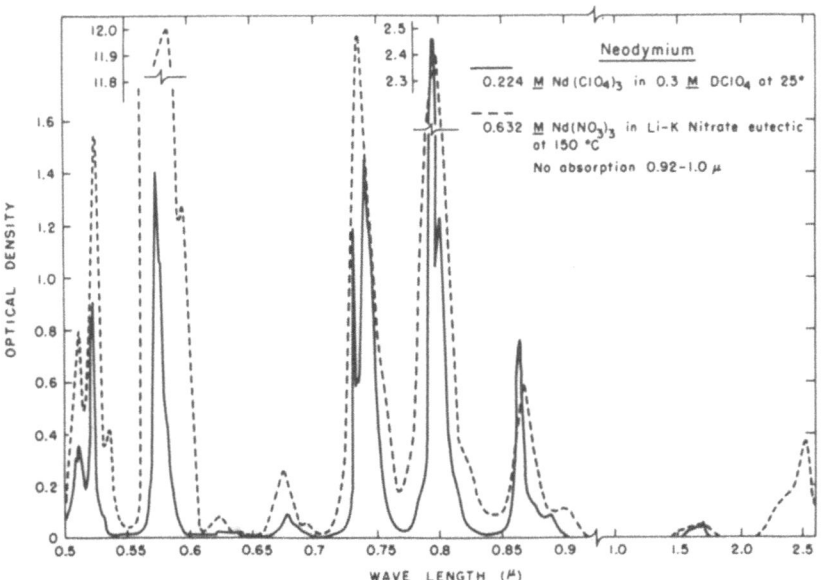

Fig. 3. Absorption spectrum of $Nd^{3+}$ in dilute $DClO_4$ at 25°C, and in molten $LiNO_3 - KNO_3$ eutectic at 150°C.

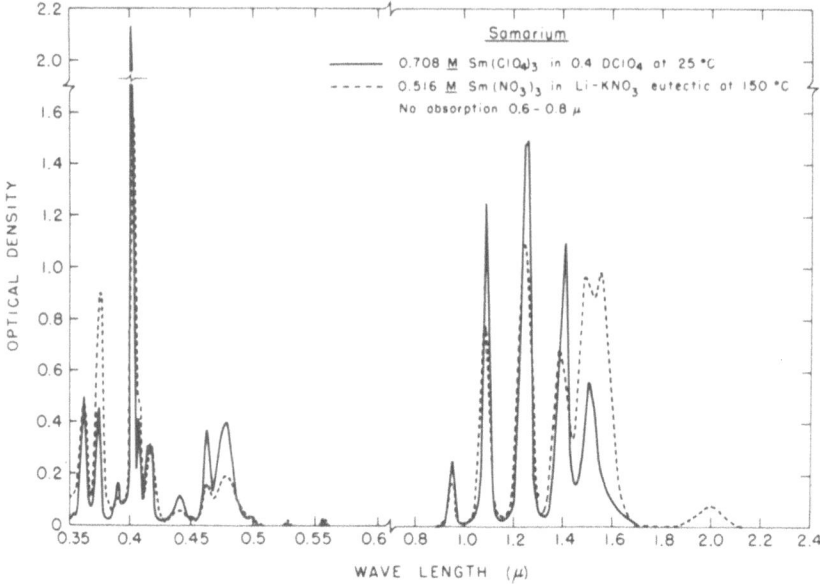

Fig. 4. Absorption spectrum of $Sm^{3+}$ in dilute $DClO_4$ at 25°C, and in molten $LiNO_3-KNO_3$ eutectic at 150°C.

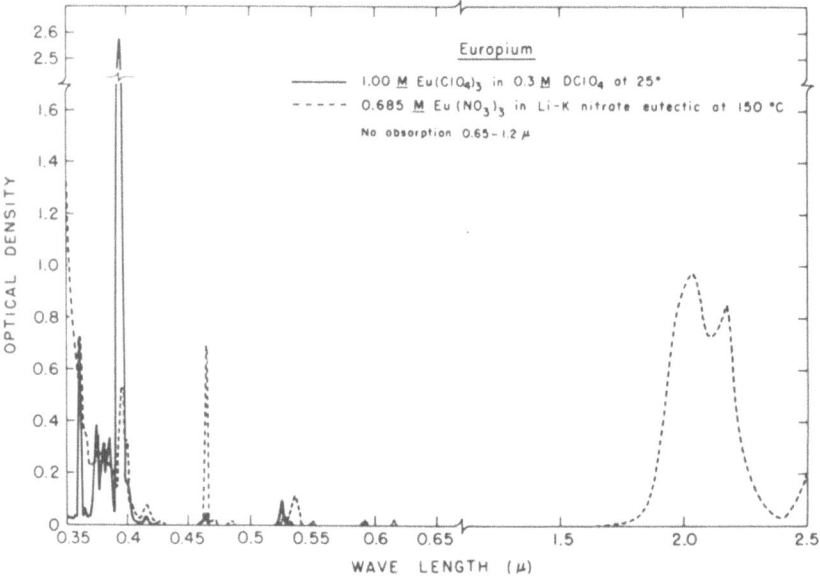

Fig. 5. Absorption spectrum of $Eu^{3+}$ in dilute $DClO_4$ at 25°C, and in molten $LiNO_3-KNO_3$ eutectic at 150°C.

## TABLE I

Comparison of Extinction Coefficients in Dilute DClO₄ at 25°C, and in Molten LiNO₃ – KNO₃ Eutectic at 150°C

| Praseodymium | | | | Neodymium | | | | Samarium | | | |
| DClO₄ | | Nitrate eutectic | | DClO₄ | | Nitrate eutectic | | DClO₄ | | Nitrate eutectic | |
| $\mu$ | $\epsilon$ | $\mu$ | $\epsilon$ | $\mu$ | $\epsilon$ | $\mu$ | $\epsilon$ | $\mu$ | $\epsilon$ | $\mu$ | $\epsilon$ |
|---|---|---|---|---|---|---|---|---|---|---|---|
|  |  | 1.925 | 2.45 |  |  | 2.512 | 0.54 |  |  | 2.0 | 0.15 |
|  |  |  |  |  |  |  |  | 1.510 | 0.79 | 1.557 | 1.90 |
| 1.550 | 4.28 | 1.538 | 3.00 | 0.794 | 11.0 | 0.798 | 3.80 | 1.248 | 2.09 | 1.246 | 2.10 |
| 1.452 | 3.44 | 1.445 | 1.73 | 0.732 | 5.31 | 0.735 | 3.20 |  |  |  |  |
| 0.469 | 4.28 | 0.468 | 1.42 | 0.575 | 6.26 | 0.582 | 19.0 | 0.402 | 3.01 | 0.404 | 3.10 |
| 0.444 | 9.98 | 0.445 | 2.46 | 0.522 | 4.06 | 0.525 | 2.40 | 0.375 | 0.64 | 0.376 | 1.74 |

| Europium | | | | Americium | | | | Curium | | | |
| DClO₄ | | Nitrate eutectic | | DClO₄ | | Nitrate eutectic | | DClO₄ | | Nitrate eutectic | |
| $\mu$ | $\epsilon$ | $\mu$ | $\epsilon$ | $\mu$ | $\epsilon$ | $\mu$ | $\epsilon$ | $\mu$ | $\epsilon$ | $\mu$ | $\epsilon$ |
|---|---|---|---|---|---|---|---|---|---|---|---|
|  |  | 2.167 | 1.21 |  |  | 1.90 | 14.2 |  |  |  |  |
|  |  | 2.030 | 1.40 | 0.813 | 66.3 | 0.796 | 36.8 |  |  |  |  |
| 0.465 | 0.05 | 0.465 | 0.94 | 0.503 | 370 | 0.509 | 110 | 0.454 | 3.8 | 0.457 | 3.1 |
| 0.394 | 2.55 | 0.395 | 0.76 | 0.455 | 1.5 | 0.458 | 17.5 | 0.433 | 6.3 | 0.434 | 4.4 |
| 0.362 | 0.71 | 0.362 | 1.11 | 0.378 | 18.2 | 0.378 | 12.3 | 0.396 | 53.5 | 0.399 | 20.9 |

termined by the water dipoles which coordinate through the oxygen atoms. In the anhydrous nitrate system, the field is again determined by a polyatomic ion coordinating through oxygen atoms to the central cation. The net effect on the central ion might be predicted to be similar, especially since the ligand-field splitting is much smaller in the lanthanides than in the outer transition series. It should be noted that one cannot deduce from the absorption spectra alone whether or not complex formation exists in these particular lanthanide – molten nitrate systems. Even in the case of strong complex formation, as established by independent measurements, the changes in lanthanide spectra as compared to those observed in dilute acid may not be large [14].

The positions of the low energy levels of the trivalent lanthanides as obtained in various crystal studies have been summarized by

Dieke and Hall [15]. For each level seen in the crystals there is an
encompassing band in solution. The relative intensity of the band
systems seen in solution in the near-infrared as compared to the
bands in the visible region is of particular interest because of their
obvious analytical significance. The fact that a solution band of at
least moderate intensity was observed to encompass each of the levels
reported in the crystal work suggests the importance of the molten
systems establishing limited regions where crystal levels would be
expected for those valence states difficult to achieve in a crystal
matrix.

## ACTINIDE SPECTRA

Investigation of the low energy levels of the trivalent actinide
ions, indeed of any of the valence states of this group of elements,
is in its first stages. Several of the actinide elements, Bk, Cf, and Es,
are at present very restricted in their availability. Of the actinides
in the first half of the $5f$ series, the elements U through Cm exhibit a valence state of three in aqueous solution. However, in the

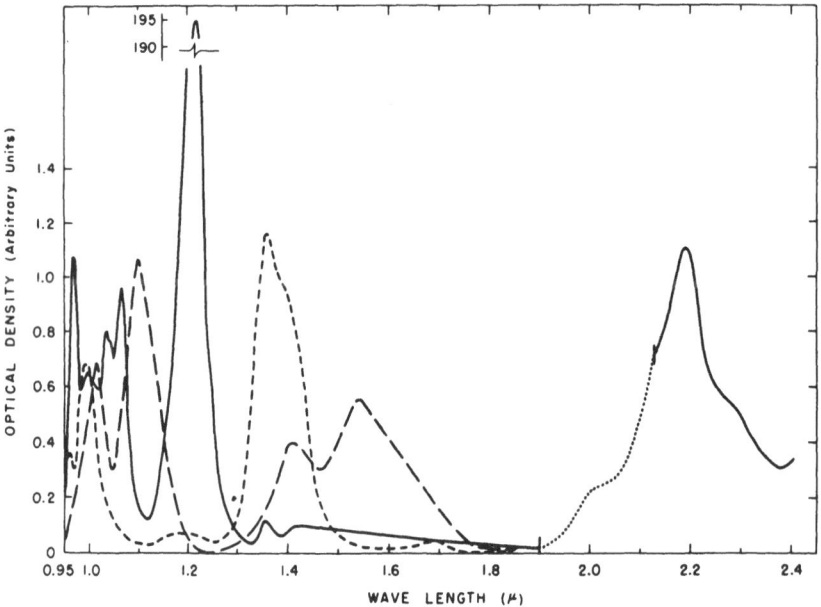

Fig. 6. The absorption spectra of $U^{3+}$ (——), $Np^{3+}$ (- - -), and $Pu^{3+}$ (— —) in dilute
$DClO_4$.

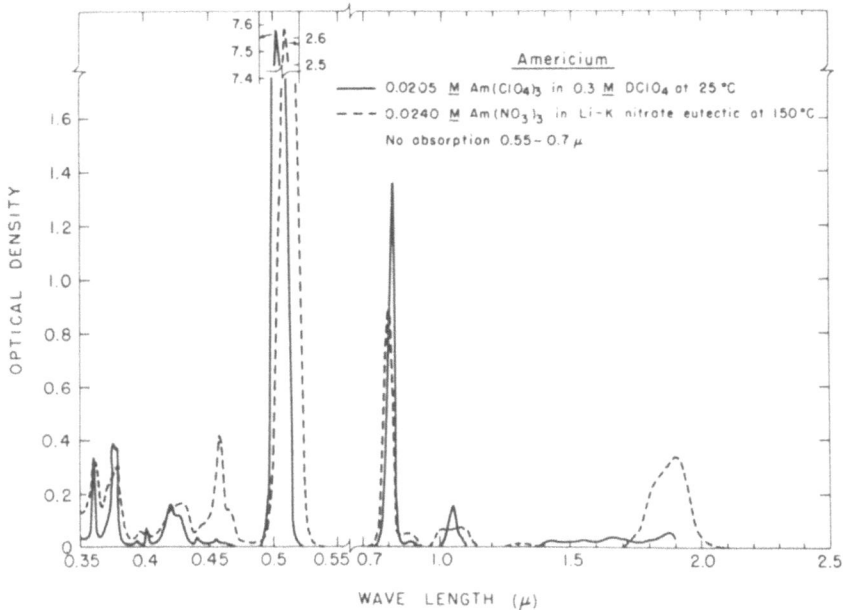

Fig. 7. Absorption spectrum of $Am^{3+}$ in dilute $DClO_4$ at 25°C, and in molten $LiNO_3 - KNO_3$ eutectic at 150°C.

nitrate melt, $U^{3+}$, $Np^{3+}$, and $Pu^{3+}$, are unstable. Each is oxidized to a higher valence state [16]. Studies in $DClO_4$ reveal the existence of bands in the near-infrared for each of these valence states, as shown in Fig. 6. Thus, these bands should also be observed in, for example, an appropriate molten chloride eutectic.

The spectra of $Am^{3+}$ and $Cm^{3+}$ in the $LiNO_3 - KNO_3$ eutectic and in dilute $DClO_4$ are shown in Figs. 7 and 8. The bands seen in $Am^{3+}$ in the near-infrared are broad and, for the most part, of relatively low intensity; however, previously unreported regions of absorption are indicated. Clearly, the molten nitrate medium has the general effect of lowering the intensity and broadening the bands; in many respects this is an extrapolation of the behavior of $Am^{3+}$ with increasing $HNO_3$ concentration [17]. In this connection it is interesting to note that in $10.0M$ $HNO_3$, electromigration studies revealed that 24% of the $Am^{3+}$ migrated to the cathode while 76% migrated to the anode [17].

Calculations for $Eu^{3+}$, similar to those made for $Pm^{3+}$, using the tabulated results of Elliott, Judd, and Runciman [11], predict

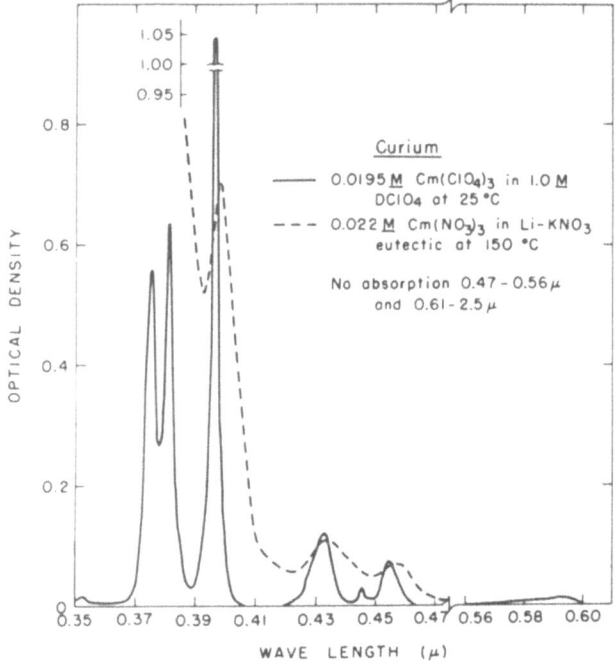

Fig. 8. Absorption spectrum of $Cm^{3+}$ in dilute $DClO_4$ at 25°C, and in molten $LiNO_3 - KNO_3$ eutectic at 150°C.

that the $^7F_6$ level will lie near 2.08 $\mu$, in agreement with experimental results. The next excited multiplet level lies at considerably higher energies. In contrast to the experimental results for $Eu^{3+}$, which is the lanthanide analogue of $Am^{3+}$, there is considerable structure observed in the near-infrared region for $Am^{3+}$; this is typical of the actinide spectra compared to their light lanthanide analogues. In both series there is a migration of bands toward the ultraviolet in progressing from $f^1$ to $f^7$ systems; however, the bands in the actinide spectra appear to be displaced toward the infrared relative to their lanthanide analogues. Thus, it is of interest to find somewhat isolated low-energy bands at essentially the same wavelength for the trivalent ions of the two inner transition series elements with the same number of $f$-electrons.

Trivalent curium with the $5f^7$ electronic structure would not be expected to have bands in the near-infrared; in this it should be similar to $Gd^{3+}$. However, the region was investigated for any evidence of possible wide splitting of the $^6P$ level. No bands were found beyond those observed in aqueous solution. A study of the absorption spectrum of $Cm^{3+}$ in strong $HNO_3$ solutions was severely limited by species arising from the radiolytic decomposition of the solvent [3]. From this standpoint, there was no problem in the molten nitrate melt.

The theoretical treatment of the absorption spectra of the trivalent lanthanide ions has been quite successful. In some cases the assumption of pure $(LS)$ coupling has given good agreement between the multiplet structure calculated and that obtained experimentally, especially for the low-lying levels. A better approximation involves intermediate coupling calculations, and these have now been reported in the literature for several of the trivalent lanthanides. For the lines observed in the visible region, the comparison of theory and experiment becomes very complex because of the multiplicity and overlapping of levels. The theoretical treatment of lanthanide spectra proceeds by first expressing the electrostatic interaction in terms of parameters that can be evaluated to give the center of gravity of the various levels. The splitting of these levels by spin-orbit interaction is small in comparison to the electrostatic or Coulomb splitting. Finally, the ligand field gives rise to a Stark splitting of the various multiplet levels. This latter effect is small in comparison to the spin-orbit splitting. The $5f$ electrons of the actinide elements are relatively less well shielded than the $4f$ electrons of the lanthanides; this would lead one to predict that the ligand-field effects would be relatively much larger. The treatment of this problem by Satten and coworkers [18], in the case of $U^{4+}$ $(f^2)$, would place the first (or possibly the second) Stark level at 1000–5000 $cm^{-1}$ above the ground level. In comparison, the first Stark level in $Pr^{3+}$ $(f^2)$ is experimentally placed at $ca.$ 33 $cm^{-1}$ above ground [19]. If Satten's interpretation is correct, bands found in the visible and near-infrared for the actinide ions would correspond to entirely different types of transitions than those in corresponding lanthanide elements. Satten's results could, however, also be explained in terms of the particular crystal lattice he used [18]. As a result of experimental work on several different actinide elements including $U^{4+}$, Conway found that the positions of the experimentally observed lines could be calculated in a manner analogous to that for the lanthanide spectra [20,21]. Thus, the bands observed in actinide spectra are to be identified with multiplet splitting, as in the lanthanide case. Jorgensen earlier adapted the same interpretation of actinide spectra [22]. Conway has suggested that the reason for the similarities in interpretation of actinide and lanthanide spectra lies in the fact that while the effects of the ligand field and spin-orbit interaction are much larger for the $5f$ electrons than for those in the $4f$ shell, the ratio of spin-orbit splitting to that of ligand-field splitting is of the same order of magnitude in both series of elements [21]. The exception to this interpretation could come in the case of a particular crystal matrix that changes this ratio.

The investigation of the absorption spectra in molten $LiNO_3$–$KNO_3$ eutectic and in $DClO_4$ of the elements in the second half of the lanthanide series is in progress, with indications that the situation is analogous to that in the first half of the series. Study of the second half of the $5f$ series will be limited to Bk, Cf, and Es, since it appears unlikely that macro amounts of trans-Es elements can be accumulated because of their short half-lives.

## ACKNOWLEDGMENT

The help of Mr. Robert McBeth in carrying out the experimental work is gratefully acknowledged.

## REFERENCES

1. J. P. Young and J. C. White, *Anal. Chem.* **31**, 1892, 1959.
2. J. C. Sullivan, D. Cohen, and J. C. Hindman, *J. Am. Chem. Soc.* **79**, 3672, 1957.
3. W. T. Carnall and P. R. Fields, *J. Am. Chem. Soc.* **81**, 4445, 1959.
4. J. M. Florence, C. C. Allshouse, F. W. Glaze, and C. H. Hahner, *J. Res. Natl. Bur. Standards* **45**, 121, 1950.
5. G. P. Smith and C. R. Boston, *Ann. N. Y. Acad. Sci.* **79**, 930, 1960.
6. J. Greenberg and L. J. Hallgren, *Rev. Sci. Insts.* **31**, 444, 1960.
7. J. Greenberg and L. J. Hallgren, *J. Chem. Phys.* **33**, 900, 1960.
8. J. G. Conway and J. B. Gruber, *J. Inorg. Nucl. Chem.* **14**, 303, 1960.
9. D. C. Steward, *Argonne National Laboratory Report ANL-4812*, 1952.
10. J. G. Conway and J. B. Gruber, *J. Chem. Phys.* **32**, 1586, 1960.
11. J. P. Elliott, B. R. Judd, and W. A. Runciman, *Proc. Roy. Soc. (London)* **240A**, 509, 1957.
12. S. P. Cook and G. H. Dieke, *J. Chem. Phys.* **27**, 1213, 1957.
13. D. Gruen, S. Fried, P. Graf, and R. L. McBeth, *Proceedings of the Second International Conference on the Peaceful Uses of Atomic Energy* Vol. 28, Pergamon Press, New York, 1959, p. 112.
14. T. Moeller and J. C. Brantley, *J. Am. Chem. Soc.* **72**, 5447, 1950.
15. G. H. Dieke and L. A. Hall, *J. Chem. Phys.* **27**, 465, 1957.
16. D. M. Gruen, R. L. McBeth, J. Kooi, and W. T. Carnall, *Ann. N. Y. Acad. Sci.* **79**, 941, 1960.
17. G. N. Yakovlev and V. N. Kosyakov, *Proceedings of the International Confer-*

*ence on the Peaceful Uses of Atomic Energy* Vol. VII, United Nations, New York, 1956, p. 363.

18. R. A. Satten, D. Young, and D. M. Gruen, *J. Chem. Phys.* **33**, 1140, 1960.
19. G. H. Dieke and R. Sarup, *J. Chem. Phys.* **29**, 741, 1958.
20. J. G. Conway, *J. Chem. Phys.* **31**, 1002, 1959.
21. J. G. Conway, Private communication.
22. C. K. Jorgensen, *Kgl. Danske Videnskab. Selskab*, Mat-fyz. Medd. **29**, No. 11, 1955.

# Isotopic Substitution in the Analysis of the Infrared Spectra of High Polymers

S. Krimm

University of Michigan
Ann Arbor, Michigan

One of the important methods for identifying the origin of bands in the infrared spectrum of a high polymer, and therefore aiding in their assignment, is isotopic substitution. The most useful of such substitutions is that of deuterium for hydrogen, which leads to a lowering of the frequencies of all modes that involve the motions of the substituted hydrogen atoms. In the past this effect has only been discussed qualitatively. The use of a recently proposed approximate isotopic frequency rule, however, now makes it possible in many cases to identify the particular normal mode associated with a given band from the magnitude of its frequency shift on deuteration. This can be an important aid in the analysis of a spectrum. The applications and limitations of this rule, as well as the general problems associated with the study of deuterated polymers, will be discussed with reference to several polymers, such as polyethylene, polystyrene, and polyvinyl chloride.

# Infrared Spectrophotometric Differences Between Some Substituted Anilines and Their Hydrochlorides

## Sister Miriam Michael Stimson, O. P.

Siena Heights College
Adrian, Michigan

The ortho- and paraphenylenediamines, aminophenols, and their hydro-chlorides and para-anisidine and the hydrochlorides of ortho- and paraanisidine were studied primarily in the regions $2.7-4\mu$ and $5.7-8\mu$. All the materials were prepared in KBr disks in concentrations of the order of $1-2 \cdot 10^{-6}M$. Ordinate expansion was used to give full-scale deflection. The order of the extent of hydrogen bonding, as determined by the shift in the N-H stretching frequencies as well as C-N absorption, is anisidine < phenylenediamine < aminophenol. The aromatic vibrations are modified in the same sequence.

Orthoaminophenol does not show the O–H stretching frequency nor the usual phenolic character in the $8-\mu$ region for the free base. Both of these characteristics reappear on formation of the hydrochloride.

# The Vibrational Spectrum and Structure of Symmetric Tetrabromoethane and Its Deuterate Analogs

E. A. Piotrowski, S. Sundaram, S. I. Miller, and F. F. Cleveland

Illinois Institute of Technology
Chicago, Illinois

Raman and infrared spectral studies have indicated the existence of rotational isomerism for substituted ethanes. Tetrabromoethane (CBr$_2$H-CBr$_2$H) is considered to exist in the *trans* and *gauche* forms. Therefore, the analysis of the Raman and infrared spectra entails the assignment of the vibrational frequencies to one or the other form. A partial assignment of this type has been made by Kagarise [1] for C$_2$H$_2$Br$_4$. A more complete assignment could be made if knowledge of the vibrational frequencies of C$_2$D$_2$Br$_4$ were available. In the present investigation the Raman and infrared spectra of C$_2$H$_2$Br$_4$ and C$_2$D$_2$Br$_4$ have been obtained. As the Raman and infrared spectra of C$_2$H$_2$Br$_4$ obtained in this investigation are essentially the same as those reported by Kagarise [1], they will not be given here. As for C$_2$D$_2$Br$_4$ the Raman displacements in cycles per centimeter, relative intensities, and depolarization factors ($\Delta\sigma(1)\rho$) are: 63(15)0.73; 112(13)0.55; 146(17)0.70; 175(19)0.68; 217(100)0.11; 434(7)0.77; 520(20)0.03; 603(6)0.35; 659(25)0.43; 793(2)0.82; 835(1)0.86; 873(3)0.72; 900(16)0.65 944(0)0.66?; 973(1)0.86; 1065(12)0.21; and 2228(12)0.34. The wave numbers in cycles per centimeter for the more prominent bands in the infrared spectrum of C$_2$D$_2$Br$_4$ are: 430(s); 520(s); 569(m); 592(vs); 611(m); 656(vs); 833(vs); 900(vs); 976(vs); 1069(s); and 2232(vs). The Raman and infrared data for C$_2$H$_2$Br$_4$ and C$_2$D$_2$Br$_4$ have been compared to arrive at a complete assignment of the observed bands. As a check on the assignments, a normal coordinate treatment using the Wilson FG matrix technique has been carried out for the *trans* isomer of each molecule.

1.  R. E. Kagarise, *J. Chem. Phys.* **24**, 300, 1956.

# Normal Coordinate Treatment of Biphosphine

## J. S. Ziomek and G. C. Madapallimatam

De Paul University
Chicago, Illinois

Infrared and Raman spectra for biphosphine and deuterobiphosphine were collected and examined for the wave numbers, intensities, and depolarization factors. Also, the frequency assignments were examined on the basis of the $C_2$ model. Then a normal coordinate treatment (FG matrix method) was conducted for this model. F matrix elements adopted from $PH_3$ and $PD_3$ were used as a first approximation and then adjusted to yield-calculated wave numbers in good agreement with experiment.

# Phosphorescence Spectroscopy as a Research Tool in Organic Chemistry

O. W. Adams

Armour Research Foundation
Chicago, Illinois

Some of the many interesting and important fields in which phosphorescence spectroscopy has been applied in the past or to which it might be applied in the future are briefly considered. This discussion is in reality concerned not so much with the details of phosphorescence spectroscopy itself, as with the chemical species or physical phenomena which can be investigated with the acid of this tool. Such a discussion seems pertinent at the moment not because this field of spectroscopy is new in concept, but because it appears that commercial phosphorescence spectrometers are becoming available and, hence, their potential usefulness in various areas should be recognized.

# The Fluorescence of Aromatic Nitrogen Heterocyclics

## B. L. Van Duuran

New York University, Medical Center
New York

Earlier reports from this laboratory revealed striking solvent effects on the positions of the fluorescence emission maxima in the indole series. The present report describes the extension of this work to the study of the fluorescence of quinoline, acridine, and other benzoquinolines in various solvents. The effects of solvents on the fluorescence spectra are compared with the UV absorption spectra of the same compounds in various solvents. These solvent effects are discussed in relation to hydrogen bonding effects, basicity, and electronic structure.

# The Analysis of Film Surfaces by Abrasion with KBr

## W. T. M. Johnson

E. I. Du Pont de Nemours & Co.
Philadelphia, Pennsylvania

We have discovered a way to determine the chemical composition of the surface of a film. This method allows determination of the upper 50 A (estimated) of film surfaces.

The method involves the light polishing of the film surface with potassium bromide powder. The powder becomes contaminated with surface material and, after collection and pressing into a disc, yields the infrared spectrum of the film surface. This spectrum gives the chemical composition of the surface.

Our results indicate that surfaces are unique; they differ chemically from bulk films. In a film containing a silicone oil, the silicone oil was found at the film surface. In polyethylene film containing an amide slip agent, the amide was found at the film surface. Neither of these surface materials could be detected by infrared analysis of the total films.

Because many important film properties are surface properties, the ability to analyze film surfaces should be of considerable value in understanding the behavior of films.

# Note

In addition to the foregoing, the Symposium program included a paper by T. F. Young (University of Chicago) on *The Application of Raman Spectroscopy in the Study of Solutions of Electrolytes,* as well as three papers on Electron Paramagnetic Resonance and Nuclear Magnetic Resonance, which are abstracted on the following pages.

# Paramagnetic Resonance and Photomagnetism

## C. A. Hutchison

University of Chicago
Chicago, Illinois

In the mechansim proposed by Lewis for the phosphorescence of organic molecules, the luminescence arises from slow emission from the lowest triplet state to the ground singlet state. Because the triplet states are paramagnetic, photomagnetism is a necessary consequence of this mechanism. The investigation of paramagnetic resonance in organic phosphors permits a very detailed study of this photomagnetism. The fine structure and hyperfine structure of paramagnetic resonance spectra in these systems have permitted the identification of the triplet states, the determination of the distribution of electron spin densities in the triplet states, and the gathering of detailed structural information on the orientation of phosphorescing molecules in host crystals.

# Chemical and Analytical Applications of High-Resolution NMR

L. Johnson

Varian Associates
Palo Alto, California

---

An almost unique feature of NMR spectra is the direct proportionality between the size of the absorption signals and the number of nuclei producing them. As a consequence of this, quantitative information can be obtained directly from an integration of the absorption signals. Several examples illustrate the use of electronic integration in obtaining quantitative information from the spectra of some organic molecules.

Molecular structure determinations are based on the measurement of chemical shifts and spin-spin couplings. Tables of typical spin-spin couplings and of chemical shifts produced by various functional groupings will be discussed. Examples of structure determinations will be presented and discussed. The general theory of NMR as related to the above topics will introduce the subject matter.

259

# Recent Advances in High-Resolution NMR Instrumentation

## S. Armstrong

Varian Associates
Palo Alto, California

---

Varian Associates has developed a unique method of effectively locking the magnetic field $H_0$ to the rotating magnetic field $H_1$.

This is done by using a nominal 5-kc frequency to modulate the modulation coils in the sample region. The effect of the 5-kc modulation of the $H_0$ fields is to introduce a second rotating field that produces sidebands 5 kc above and below the 60-Mc fundamental frequency.

Then the $H_0$ field intensity is adjusted to a value which, according to the Larmor equation, will cause the nuclei to precess at the upper sideband frequency.

The 60-Mc plus the nominal 5-kc sideband frequency appear together with the 60-Mc fundamental frequency in the control-sample (water) excitation/receiver coil. Diode–detected at the receiver output, the 5-kc beat frequency between the 60-Mc fundamental and upper sideband is so phased that if amplified and applied to the $H_0$ field modulation coils in the probe, it causes oscillation in a control loop comprised of the modulation coils, the control sample, the control receiver, and the modulating amplifier. The frequency of oscillation automatically adjusts to a value necessary to create an upper sideband frequency (60 Mc + 5 kc) which satisfies the Larmor equation for the fundamental $H_0$ field value for protons. Thus, if either the $H_0$ field or the frequency of the $H_1$ field drifts, automatic correction takes place to reestablish the upper sideband at the precession frequency of the nuclei for the mean value of $H_0$ field intensity at the samples.